Food Waste Valorization: Preparation, Properties, and Application of the Synergistic Soil Conditioners

Jiaqi Hou · Mingxiao Li · Beidou Xi · Xuan Jia ·
Ning Jiang · Wangmi Chen

Food Waste Valorization: Preparation, Properties, and Application of the Synergistic Soil Conditioners

A Book on Food Waste from Fermentation to Application

Springer

Jiaqi Hou
Chinese Research Academy
of Environmental Sciences
Beijing, China

Mingxiao Li
Chinese Research Academy
of Environmental Sciences
Beijing, China

Beidou Xi
Chinese Research Academy
of Environmental Sciences
Beijing, China

Xuan Jia
Beijing Technology and Business
University
Beijing, China

Ning Jiang
Chinese Research Academy
of Environmental Sciences
Beijing, China

Wangmi Chen
Chinese Research Academy
of Environmental Sciences
Beijing, China

ISBN 978-981-95-5361-7 ISBN 978-981-95-5362-4 (eBook)
https://doi.org/10.1007/978-981-95-5362-4

This Springer imprint is published by the registered company Springer Nature Singapore Pte Ltd.
The registered company address is: 152 Beach Road, #21-01/04 Gateway East, Singapore 189721, Singapore

To all who turn waste into worth and ideas into impact.

Preface

With the rapid economic development and accelerated urbanization worldwide, the generation of food waste has been increasing substantially each year. For instance, China produced approximately 127 million tons of food waste in 2021, accounting for over half of all municipal solid waste. The treatment of such large quantities of food waste poses a significant challenge for countries around the globe. Food waste possesses a dual nature, being both a "hazard" and a "resource". Aerobic fermentation has emerged as one of the most effective methods for treating food waste. This process efficiently utilizes the organic content within the waste, converting it into organic fertilizers or soil conditioners. This approach not only helps mitigate environmental pollution but also promotes resource recovery and valorization. This book presents our recent research advances in the preparation and application of soil conditioners derived from aerobic fermentation of food waste. It specifically focuses on the transformation of organic components and microbial microenvironment dynamics during the aerobic fermentation process, as well as the mechanisms through which a novel synergistic and high-efficiency soil conditioner enhances humification in impoverished soils and ameliorates secondary salinization.

In this book, we: (1) provide an overview of the current status of food waste generation and its valorization, along with the properties and applications of conventional soil conditioners; (2) discuss the transformation of organic components and changes in microbial microenvironments during different aerobic fermentation processes of food waste; (3) describe the preparation and characteristics of a novel synergistic and high-efficiency soil conditioner; (4) discuss the role of this conditioner in promoting humification in impoverished soils, examining its effects on humus formation, long-term dynamics of active carbon pools, and microbial community succession across various soil layers; and (5) explore its efficacy in remediating secondary salinized soils, with emphasis on improvements in soil physicochemical properties, salt migration and transformation, and interactions with soil microbial communities.

This book encompasses the current trends in food waste valorization, as well as the preparation, properties, and application of an advanced soil conditioner—produced via aerobic fermentation of food waste—for the rehabilitation of impoverished and secondary salinized soils. It serves as a valuable reference for researchers, students, industry professionals, and government officials engaged in the field of waste treatment and sustainable resource management.

Due to the limited level of editors, there are unavoidable omissions in the book so please oblige us with your valuable comments.

Beijing, China

Jiaqi Hou
Mingxiao Li
Beidou Xi
Xuan Jia
Ning Jiang
Wangmi Chen

Acknowledgements This book was financially supported by National Key Research and Development Program of China (No. 2019YFD1100304-02 and No. 2020YFD1100301-04) and the National Natural Science Foundation of China (No. 51908524).

Competing Interests The authors have no competing interests to declare that are relevant to the content of this manuscript.

Contents

List of Contributors

This book is authored by Jiaqi Hou, Mingxiao Li, Beidou Xi, Xuan Jia, Ning Jiang, and Wangmi Chen. Specifically, Chap. 1 is authored by Ning Jiang, Chap. 2 by Xuan Jia, Chaps. 3 and 4 by Jiaqi Hou, and Chap. 5 by Wangmi Chen. The manuscript was revised by Jiaqi Hou and Ning Jiang and reviewed and finalized by Beidou Xi and Mingxiao Li.

Abbreviations

AD	Anaerobic digestion
AK	Available potassium
AP	Available phosphorus
BOF	Bio-organic fertilizer
CCA	Canonical correspondence analysis
CK	Conventional fermentation
CPMAS	Cross-polarization magic-angle spinning
CPMI	Carbon pool management index
DCA	Detrended correspondence analysis
EC	Electrical conductivity
EEM	Excitation-emission matrix
EOC	Easily oxidizable organic carbon
FRI	Fluorescence regional integration
FTIR	Fourier-transform infrared spectroscopy
GMD	Geometric mean diameter
HF	Chemical fertilizer
HFOC	Heavy fraction organic carbon
HTC	Hydrothermal carbonization
JJ	Fermentation with phased inoculation
LFOC	Light fraction organic carbon
LMWOC	Low-molecular-weight organic carbon
LOC	Labile organic carbon
MBC	Microbial biomass carbon
MBN	Microbial biomass nitrogen
MW	Molecular weight
MWD	Mean weight diameter
OTUs	Operational taxonomic units
PARAFAC	Parallel Factor Analysis
PHA	Polyhydroxyalkanoates
POC	Particulate organic carbon
RC	Rapid hydrothermal composting

RDA	Redundancy analysis
ROC	Recalcitrant organic carbon
SCP	Single-cell protein
SHF	Synergistic and high-efficiency conditioner
SMPs	Soluble microbial products
SOC	Soluble organic carbon
SOM	Soil organic matter
TC	Total carbon
TDS	Total dissolved solids
TF	Traditional organic fertilizer
TN	Total nitrogen
TOC	Total organic carbon
TP	Total phosphorus
UV-Vis	Ultraviolet-visible
VFA	Volatile fatty acid
WSC	Water-soluble carbon

Chapter 1
Introduction

Abstract Food waste primarily originates from food processing, catering services, and household activities, and can be categorized into restaurant waste, kitchen waste, and waste edible oils. It is characterized by high moisture content, rich organic matter, as well as the presence of oils, salts, and various nutrients. These properties make it prone to spoilage and environmental issues, while also endowing it with significant potential for resource recovery. Through appropriate technologies, it can be converted into energy, fertilizers, and high-value-added products. Valorization techniques mainly include aerobic composting, anaerobic digestion, animal feed production, thermal treatment, and high-value utilization technologies such as biodiesel, bioplastics, feed protein, and biofuel production. Among these, aerobic composting and anaerobic digestion are relatively mature and widely applied, while most of the high-value technologies remain at the laboratory or pilot stage, requiring further research for industrial-scale implementation. Through aerobic fermentation, food waste can be transformed into soil conditioners for soil improvement. Soil conditioners significantly increase soil organic matter content, improve soil structure, and increase microbial activity and diversity, thereby promoting soil health and crop growth. They demonstrate remarkable effectiveness in boosting soil fertility and regulating microbial communities, outperforming the application of chemical fertilizers alone. Therefore, the valorization of food waste is of great significance for reducing pollution and carbon emissions and advancing the circular economy. The products derived from its aerobic fermentation hold considerable promise for soil remediation and sustainable agricultural production.

1.1 Types and Characteristics of Food Waste

1.1.1 Definition and Types of Food Waste

With the rapid development of economies and the advancement of urbanization, the generation of food waste has been increasing rapidly year after year. Its growing impact on daily life and the living environment has made it a pressing challenge for

J. Hou et al., *Food Waste Valorization: Preparation, Properties, and Application of the Synergistic Soil Conditioners*, https://doi.org/10.1007/978-981-95-5362-4_1

countries worldwide. Food waste refers to various waste materials generated from residential daily life, as well as from activities such as food processing, food service, and institutional catering. It primarily falls into three categories: (1) Catering Waste: Leftovers and food scraps from establishments like hotels, restaurants, and canteens. (2) Kitchen Waste: Uncooked food scraps from the kitchens of restaurants, canteens, and residential homes. (3) Waste Edible Oils: Waste animal and vegetable oils, and wastewater containing these oils, from food service, food production and processing, and on-site food preparation and sales (Pan et al. 2024).

1.1.2 *Characteristics of Food Waste*

The composition, properties, and volume of food waste are significantly influenced by factors such as geographical region, season, residents' dietary habits, and the methods used for storage and collection. As shown in Table 1.1, the primary chemical components of food waste in some Chinese cities include starch, cellulose, protein, fat, and inorganic salts (Deng 2019). From the data in the table, the main characteristics of food waste in China are: (1) High moisture content, typically ranging from 70 to 85%. (2) High organic matter content, with components like protein and fat accounting for 80–93% (on a dry basis), making it well-suited for microbial treatment. (3) Rich in fats and oils, with a content of approximately 2–3%, which can be recovered and processed into resources like biodiesel. (4) High salt content, around 1%, which can affect its subsequent use as an organic fertilizer. Additionally, it contains small amounts of nutritive elements such as P, K, Na, Ca, and Fe.

The data in Table 1.1 confirms that food waste is characterized by high moisture and organic matter content. These properties cause it to spoil and decompose easily, emit foul odors, and breed pathogenic microorganisms. It can also spread bacteria and viruses, posing a significant threat to environmental sanitation. Besides, food waste is rich in balanced nutrients, making it an excellent raw material for composting, fermentation, and animal feed production. The resulting products are high in organic matter and various trace elements, offering significant economic and ecological value

Table 1.1 Main composition of food waste in selected Chinese cities (%)

City	Moisture content	Organic matter[a]	Crude protein[a]	Crude fat[a]	Oil Content[b]	Salt content[b]
Beijing	74.34	80.21	25.86	24.77	3.12	0.36
Tianjin	70.99	85.64	24.30	25.96	2.63	0.70
Chongqing	85.07	92.66	14.45	17.02	1.96	0.24
Suzhou	84.43	82.98	21.80	29.30	3.28	0.70
Hangzhou	74.94	91.50	16.46	24.31	2.09	1.32

[a] On a dry basis
[b] On a wet basis

(Xin et al. 2023). Therefore, food waste possesses a dual nature, being both a "hazard" and a "resource". It is a "resource-type" waste. Implementing synergistic treatment for pollution and carbon reduction can generate considerable social, economic, and environmental benefits, contributing to the goals of "peak carbon emissions" and "carbon neutrality". Conversely, improper disposal can cause severe harm to the ecosystem and human health.

1.2 Current Status of Food Waste Valorization

1.2.1 International Regulations and Policies on Food Waste Treatment

Food waste is a major component of municipal solid waste. Globally, approximately 50 billion tons of municipal solid waste are generated each year, with food waste accounting for 10–20%. Regulations and policies on food waste treatment vary among countries, differing based on region, economic level, and environmental philosophies.

The European Union's 1999 Landfill Directive explicitly prohibits the direct landfilling of biodegradable waste and encourages valorization methods such as composting and anaerobic digestion. The 2003 Animal By-Products Regulations prohibits the use of food waste for animal feed production and encourages alternatives to landfilling, such as recycling and composting. Countries like Germany and the United Kingdom widely use anaerobic digestion technology to convert food waste into biogas for power generation. The Swedish government provides dedicated collection facilities to recover food waste for production into biofuel (for vehicles) or organic fertilizer.

In the United States, the federal Resource Conservation and Recovery Act (RCRA) regulates solid waste management but does not mandate the separation of food waste. The 2009 National Green Building Standard encourages the installation of kitchen food waste disposers in new residential buildings to reduce landfilling. At the state level, Vermont passed food waste legislation in 2012, requiring all residents and businesses in the state to progressively reduce organic waste, with the goal of banning all food scraps from landfills by 2020. California mandates organic waste recycling, promoting composting and anaerobic digestion, while Pennsylvania strongly promotes waste-to-energy and composting projects for food waste. Overall, although a large amount of food waste is still landfilled, policy is gradually shifting towards valorization.

In its Fourth Fundamental Plan for Establishing a Sound Material-Cycle Society, Japan's Ministry of the Environment further classified food waste as a "biological resource." The plan stipulates that food waste should be thoroughly converted into animal feed, fertilizer, or used for energy recovery. For food waste that is difficult to process into feed or fertilizer, it can be utilized as a raw material for biopower generation, heat recovery, and other applications.

In the past, South Korea's primary method for handling food waste was landfilling. However, with the enactment of laws such as the Waste Control Act and the Act on the Promotion of Saving and Recycling of Resources, South Korea's treatment methods in recent years have shifted to anaerobic digestion with biogas recovery and aerobic treatment of sludge in bioreactors, achieving positive results.

1.2.2 Relevant Policies and Regulations in China

In 2021, China's food waste production reached 127 million tons, accounting for over 50% of the total municipal solid waste. The valorization of such a vast amount of food waste presents a significant challenge for the country in its urbanization process.

China places great importance on the handling and disposal of food waste. Starting in 2010, the nation launched pilot programs for the valorization and harmless treatment of food waste. By the end of 2021, a total of 79 pilot cities had passed inspection and been approved. In 2020, the State Council issued the Law of the People's Republic of China on the Prevention and Control of Environmental Pollution by Solid Waste (Revised Draft). This law mandates the establishment of a food waste management system, requiring that entities generating and collecting food waste hand it over to qualified professional units for harmless treatment. It also prohibits livestock and poultry farms and breeding communities from using untreated food waste as feed. In 2021, the National Development and Reform Commission released the Notice on the Action Plan for Deepening the Price Mechanism Reform during the 14th Five-Year Plan Period, which promotes a metered charging system for non-residential food waste. On February 9, 2024, the State Council issued the Opinions of the General Office of the State Council on Accelerating the Construction of a Waste Recycling System. This document calls for the orderly advancement of food waste treatment facility construction and for enhancing the level of energy and resource recovery from waste cooking oil and other food waste. The implementation of these relevant laws and regulations has advanced the progress of food waste valorization in China.

1.2.3 Food Waste Valorization Technologies

Food waste treatment technologies are categorized into three main types: landfill and incineration, biological treatment, and feed utilization. Since the resource recovery rate for landfill and incineration is virtually zero, biological treatment and feed utilization are currently the most widely applied technologies for food waste valorization. Among these, biological treatment technologies include aerobic fermentation to produce biomass fertilizer and anaerobic digestion to generate clean biogas energy.

Furthermore, in recent years, researchers have developed novel technologies for the valorization of food waste, such as the valorization of waste oils derived from food waste, the production of feed protein and energy recovery from food waste (Chang et al. 2021).

1.2.3.1 Aerobic Composting Technology

Aerobic composting refers to the biological degradation of organic matter piled on the ground or in specialized fermentation units by aerobic microorganisms under oxygenated conditions, ultimately forming stable humus with high fertility. Food waste is well-suited for composting due to its high organic matter content, comprehensive nutrient elements, and low carbon-to-nitrogen ratio (C/N), which make it an excellent substrate for microorganisms. Key techniques include traditional aerobic composting, hydrothermal-enhanced fermentation, and inoculation-enhanced fermentation. Building on aerobic composting, earthworms can also be introduced. Utilizing their own rich enzyme systems, earthworms convert the organic matter in food waste into nutrients that are easily usable by themselves or other organisms, thereby accelerating the stabilization process of the compost. Taking inoculation-enhanced fermentation technology as an example, high-temperature composite microbial strains with strong vitality and proliferation capabilities are selected from nature. In biochemical treatment equipment, these strains facilitate the rapid high-temperature aerobic fermentation of food waste and other organic waste, quickly degrading and converting various organic substances into bio-humic acid fertilizer. This humic acid fertilizer can be used as a soil conditioner to improve soil quality, which helps reduce the frequency of chemical fertilizer use, increase crop yields, and enhance the quality of agricultural products. The advantages of this technology include a fast conversion speed, high organic matter utilization rate, high product consistency, and the ability to enter industrial product sales channels. This technology is the mainstream representative process for aerobic fermentation and has been successfully applied in cities such as Beijing, Guangzhou, Chengdu, and Urumqi. However, a drawback is that the liquid phase from this process enters the sewage treatment system, which increases the wastewater load and results in the loss of organic matter from the liquid phase.

1.2.3.2 Anaerobic Fermentation Technology

Anaerobic digestion is a process where, in the absence of oxygen, facultative and anaerobic microorganisms break down complex organic matter into small-molecule organic substances and inorganic matter. Energy substances like methane and hydrogen can be produced during this process. Furthermore, various organic acids and alcohols, such as ethanol, acetic acid, butyric acid, glucoamylase, and lactic acid, can be obtained through anaerobic digestion, thereby achieving volume reduction, mass reduction, and valorization of food waste. The biogas produced from anaerobic digestion can be used as fuel gas, and the digestate from the anaerobic digester can undergo a secondary fermentation to produce fertilizer. Typically, the methane content in biogas from anaerobic digestion is 60–75%. According to the experience from the first phase of the Hangzhou food waste treatment project, with a treatment capacity of 200 t d^{-1}, biogas production can reach 13,000 m d^{-1}. When the methane concentration in the biogas is 60%, it can generate approximately 26,000 (kW h) d^{-1} of electricity, and the oil and grease recovery rate can reach 88%. The biogas produced from anaerobic digestion can also be purified and pressurized before being fed into the natural gas grid for residential daily use.

1.2.3.3 Feed Utilization Technology

Food waste contains a large amount of organic nutrients, giving it a considerable advantage for conversion into animal feed. Currently, there are two main types of feed utilization technologies: dry feed and protein feed. The production of "dry feed" requires the material to be dried at 95–120 °C for at least 2 h to achieve a moisture content of less than 15% and an impurity level below 5%. "Protein feed" is composed of the microorganisms themselves and their protein secretions and is dried at 60–80 °C. Due to the wide variety of sources and complex composition of food waste, there are many safety risks associated with its use as feed, such as "bio-homology", pathogens, heavy metals, and toxic organic compounds. To improve the reliability and safety of the food waste-to-feed process, the Technical Code for Food Waste Treatment (CJJ 184—2012) specifies the requirements for the food waste used for feed production and the corresponding production conditions.

1.2.3.4 Thermal Treatment Technologies

Thermal treatment technologies primarily include pyrolysis and hydrothermal carbonization (HTC). Due to its typically high moisture content (often exceeding 90%), food waste is most commonly treated using Hydrothermal Carbonization (HTC) technology. HTC is a process where wet biomass (carbohydrates) serves as the feedstock in a closed system. Under elevated temperatures and self-generated pressure, it undergoes a series of complex reactions to be converted into carbonaceous materials.

HTC is a technology capable of achieving the harmless treatment of food waste and has recently become a focus of research, also in the context of sludge treatment. Compared to traditional pyrolysis, the hydrothermal method is relatively milder. Solid hydrochar can be obtained via solid–liquid separation, and the process has lower equipment requirements. Crucially, HTC does not require energy-intensive drying pre-treatment, achieving carbonization in a single step, which makes it more suitable for industrial application.

The ability of HTC to process wet feedstock makes it particularly attractive for treating high-moisture content food waste. Furthermore, since the reactions occur in water, there is significant potential for tailoring the process. By selecting appropriate additives, other elements can be introduced into the carbonaceous product, enabling the synthesis of carbon composites with different morphologies, or allowing for elemental doping and surface modification. This versatility opens up vast possibilities for producing a wide variety of value-added products. However, current research on HTC for food waste is predominantly confined to laboratory and pilot scales. Applied research focused on the engineering and commercialization of hydrothermal carbonization for food waste treatment requires further significant development.

1.2.3.5 Valorization Technologies for Waste Oil from Food Waste

(1) Biodiesel Production from Waste Oil

Utilizing waste oil/fat as a raw material for biodiesel production represents a significant approach to the valorization of lipids derived from food waste, constituting one of the most concentrated areas of waste oil valorization. Biodiesel refers to long-chain fatty acid mono-alkyl esters obtained through specific chemical reaction processes using feedstocks such as vegetable oils, animal fats, and waste cooking oils. Characterized by non-toxicity, renewability, and biodegradability, biodiesel serves as a green alternative energy source capable of replacing traditional fossil fuels. Key technologies for biodiesel production from waste oils include pyrolysis, transesterification, enzymatic catalysis, and supercritical methanol methods. However, oils derived from food waste often contain free fatty acids and various polymers that hinder the transesterification reaction. In practical production, acid or alkali catalysts must be selected based on the specific properties of the feedstock to ensure the reaction proceeds efficiently. Consequently, current research focuses predominantly on developing efficient, economical, and stable biodiesel production technologies.

(2) Bioplastic Production from Waste Oil

The production of biodegradable plastics, specifically polyhydroxyalkanoates (PHA), from waste oil/fat via microbial synthesis is an emerging field. PHA constitutes a family of polyesters entirely synthesized by microorganisms, offering advantages such as excellent thermoplasticity, biodegradability, and biocompatibility. It holds significant promise as a substitute for conventional plastics, thereby reducing the generation of "white plastic waste."

Waste oil/fat from food waste, characterized by its low market value and high carbon content, serves as an effective carbon source for microbial cultivation and has been extensively studied for bioplastic synthesis. Research has identified several microbial species capable of producing PHA from waste oils, including *Cupriavidus*, *Rhodobacter*, *Ralstonia*, *Alcaligenes*, *Pseudomonas*, and *Enterobacter*. Nevertheless, most current research on bioplastic production from waste oils remains at the laboratory scale. Significant challenges remain before the commercial-scale synthesis of PHA from waste oils can be realized. Furthermore, the impact of the complex composition of food waste-derived oils on PHA synthesis efficiency requires in-depth investigation.

1.2.3.6 Technologies for Producing Feed Protein from Food Waste

(1) Insect Farming Technology Utilizing Food Waste

The conversion of food waste into nutrients for other organisms via bioconversion technologies is a relatively recent development. Utilizing insects as a protein source presents a method for recycling food waste while producing protein-rich feed. Current research primarily focuses on employing earthworms, flies, mealworms (*Tenebrio molitor*), locusts, black soldier fly (*Hermetia illucens*), and silkworms for food waste treatment. Among these, the black soldier fly exhibits the most favorable biological characteristics. It thrives in the high-oil, high-salt environment typical of food waste, possesses a broad diet, strong stress resistance, and a suitable development cycle. Its nutritional profile is comprehensive; fully developed larvae yield biomass containing 40–45% protein and up to 35% fat (dry matter basis), making it the most industrially promising species for insect-based food waste valorization.

Insect farming technology using food waste addresses the shortage of protein feed while achieving waste reduction, harmlessness, and resource recovery, offering considerable environmental and economic benefits. However, current regulations in the European Union and the United States prohibit its use as novel livestock feed, hindering the widespread adoption of insect protein as a feed resource. On one hand, production scale is constrained by factors such as production costs, low automation levels in insect farming equipment, and safety concerns, severely limiting factory-scale and standardized production. On the other hand, the industrialization of insect protein production from food waste and market access for the resulting products necessitate supportive national legal and policy frameworks.

(2) Microbial Protein Feed Production Technology from Food Waste

Single-cell protein (SCP), also known as microbial protein or bacterial protein, primarily consists of yeast protein, bacterial protein, and algal protein. It contains 50–85% crude protein, with a complete amino acid profile and high bioavailability. SCP also contains vitamins, minerals, fats, and carbohydrates, offering superior nutritional value compared to fish meal and soybean meal. Research on producing SCP from renewable resources in China dates back to the 1980s and has achieved significant progress. Food waste contains abundant fermentable and non-fermentable sugars. A novel approach involves utilizing microorganisms such as yeasts, molds, and bacteria, based on solid-state fermentation principles, to degrade food waste and produce higher-value SCP feed products. Yeasts were among the earliest microorganisms used for SCP production; their rich protein content makes them suitable as animal feed additives to enhance nutritional value.

However, most existing research remains at the laboratory scale, with relatively few studies focusing on the practical engineering aspects of producing protein feed from food waste. Due to the complex composition of food waste, the resulting microbial protein feed may harbor pathogenic microorganisms such as foot-and-mouth disease virus or classical swine fever virus, posing potential feed safety risks.

1.2.3.7 Energy Recovery Technologies from Food Waste

(1) Biofuel Production Technology from Food Waste

Food waste contains organic compounds such as starch, sugars, fats, proteins, and cellulose. Extensive recent research confirms its suitability as a feedstock for ethanol fuel production. Common production processes include separate enzymatic hydrolysis and fermentation (SHF) and simultaneous saccharification and fermentation (SSF), with hydrolysis/saccharification and fermentation being the key control steps. Compared to ethanol, butanol offers higher heat of combustion, lower water solubility and corrosivity, and does not produce SOx or NOx emissions, while also reducing greenhouse gas emissions, making it another current research focus for biofuel applications. However, these processes have primarily been optimized under laboratory conditions. The industrialization of biofuels is currently hampered by the prevailing international oil prices.

(2) Biohydrogen Production from Food Waste via Anaerobic Fermentation

Anaerobic digestion (AD) is a crucial process for the resource recovery treatment of food waste. While substantial research focuses on methane production performance from food waste AD, improving the conversion efficiency of organic matter to methane, and optimizing fermentation process conditions, studies on the intermediate product of the acidogenesis stage—biohydrogen—remain limited. In recent years, researchers have begun investigating biohydrogen production and combined hydrogen-methane co-production from food waste to maximize energy recovery. Wang et al. studied the effects of different initial pH levels and fermentation temperatures on hydrogen production potential, intermediate metabolites, and fermentation pathways using food waste as the substrate (Wang et al. 2017). Their results indicated that controlling the fermentation temperature at 55 °C and initial pH at 6 effectively enhanced substrate utilization efficiency and hydrogen production potential of hydrogen-producing microorganisms, achieving a maximum hydrogen yield of 52.45%.

Currently, biohydrogen production via anaerobic fermentation of biomass like food waste is still confined to laboratory or pilot scales. Future research will focus on enhancing process stability by investigating fermentation technology control, parameter optimization, enrichment of efficient hydrogen-producing bacteria, and the development of high-performance anaerobic fermentation reactors for hydrogen production.

In summary, most valorization technologies for food waste and waste oil—including those for producing high-value products, feed protein, and energy—currently remain at the laboratory research stage and are still some distance from industrial-scale implementation. At present, only traditional resource recovery technologies such as aerobic composting and anaerobic fermentation have reached relative maturity and large-scale application. In particular, aerobic composting is widely adopted due to its large processing capacity, operational simplicity, high degree of harmlessness and resource recovery, low capital investment, and minimal operational and maintenance costs.

In summary, the valorization technologies for waste oils derived from food waste, the production of feed protein from food waste, and the energy recovery technologies utilizing food waste remain largely at the laboratory research stage, with considerable distance to go before achieving industrial-scale implementation. At present, only traditional resource recovery technologies—such as aerobic composting and anaerobic digestion—have reached a relatively mature level of development and large-scale application. In particular, aerobic composting is widely employed due to its high processing capacity, operational simplicity, significant degree of sanitization and resource recovery, low capital investment, and reduced operational and maintenance costs.

1.3 Properties and Application of Soil Conditioners

1.3.1 Properties and Types of Soil Conditioners

A soil conditioner is a material added to the soil to improve its physicochemical properties and biological activity. In a broad sense, organic fertilizers, inorganic fertilizers, and other chemical remediation agents are all classified as soil conditioners. Extensive research and application have shown that the application of soil conditioners can significantly enhance the stability of topsoil. At the same time, by improving soil structure and increasing soil moisture content, they can reduce the harm from soil salinity and alkalinity, thereby playing a role in soil conservation and stabilization (Cogger and Brown 2016). Currently, there is a significant amount of research both domestically and internationally on producing organic fertilizers from organic solid waste. Organic solid wastes, such as municipal solid waste and food waste, are rich in organic matter. After fermentation, they can be used as soil conditioners. This not only provides essential nutrients to the soil but also achieves a reduction in waste volume. Therefore, it is a commonly used method for preparing soil conditioners both in China and abroad. Table 1.2 lists common sources and functions of organic soil conditioners (Hou et al. 2017).

Table 1.2 Sources and functions of organic soil conditioners

Material source	Processing method	Application
Food waste	Hydrothermal hydrolysis and fermentation	Improving fertility of barren soil
Food waste, inoculation	Aerobic fermentation	Promoting the growth of beneficial soil microorganisms, enhancing agricultural sustainability
Digestate from biogas plants	High-temperature aerobic fermentation	Improving the basic fertility of agricultural soil
Biochar primarily from hickory shells	Oxygen-limited pyrolysis	Conditioning acidic soils such as fine loam and kaolin
Wood-residue biochar	High-temperature pyrolysis	Immobilization and stabilization of heavy metals
Paper, soft materials, wood chips	High-temperature pyrolysis	Improving basic soil fertility
Organic fertilizers such as wheat straw, seed cakes, and cottonseed cakes	Aerobic fermentation	Effectively preventing nitrogen mineralization, increasing the total nitrogen content in the topsoil
Cow manure	Anaerobic digestion	Ameliorating acidic soil
Chicken manure	Aerobic fermentation	Remediating cadmium-contaminated soil
Pig manure, vermicompost	Composting	Enhancing the growth of tomato seedlings
Beet pulp, municipal solid waste	Composting	Immobilization and stabilization of heavy metals

1.3.2 The Remediation Effect of Soil Conditioners on Soil

The application of exogenous nutrients (e.g., chemical/organic fertilizers, and other soil conditioners) significantly alters the proportion of beneficial elements and soil physicochemical properties. This process influences the composition of functional microbial consortia involved in carbon/nitrogen cycles and other abiotic processes, directly and indirectly regulating microbial growth and reproduction. Concurrently, it modifies the abundance and diversity of specific functional microorganisms, thereby affecting soil organic matter (SOM) sequestration capacity and cycling characteristics. Consequently, the ecological functionality of soil microbial communities serves as a critical indicator of soil fertility.

1.3.2.1 Impacts of Chemical Fertilizers

Field studies on rice cultivation demonstrate that nitrogen fertilizers elevate rhizospheric microbial biomass carbon (MBC) during identical growth stages, stimulating microbial abundance (Cai et al. 2008). Inorganic fertilizers notably improve soil microbial community structure, diversity, and abundance, though unbalanced fertilization inhibits microbial biomass nitrogen (MBN) accumulation. In wheat fields, MBN in the plow layer (0–40 cm) increases significantly with nitrogen application rates up to 240 kg ha^{-1} (Zhao et al. 2006). These findings confirm that chemical fertilizers substantially enhance soil MBN.

However, fertilizer efficacy diminishes at high application rates. Cessation of fertilization shows no impact on microbial biomass in fertile grasslands, whereas excessive nitrogen damages grass roots and reduces microbial biomass in pasture soils. Appropriate N-P application in maize fields induces negligible changes in MBC, but over-application decreases MBC/MBN while inhibiting carbon/nitrogen-fixing microorganisms (Jia 2008). In agricultural practice, intensive irrational fertilizer use disrupts SOM biochemical cycling and microbial community stability, accelerating SOM loss.

1.3.2.2 Impacts of Organic Fertilizers

Rational organic matter input enhances soil carbon sequestration capacity, improves microbial community structure/functionality, and enriches functional microorganism diversity. Organic fertilizers—rich in balanced organic compounds and trace elements—elicit priming effects on indigenous microbes through inputs of carbohydrates, proteins/peptides, amino acids, and lipids. Crucially, they compensate for soil carbon loss while invigorating microbial activity, thereby strengthening SOM cycling and nutrient assimilation. This process significantly elevates soil phosphorus, iron, manganese, and zinc availability.

Studies reveal that organic or organic-mineral fertilizers substantially increase microbial abundance, particularly in bacterial communities. Compared to chemical fertilizers or no fertilization, organic amendments significantly elevate MBC in maize fields, exhibiting positive dose–response correlations (Li et al., 2004). In incubation trials, lupine-based organic fertilizers increased microbial biomass 2- to fivefold after 91 days. Similar MBC enhancements occur in spring wheat-amaranth rotation systems. Rice paddy studies in Hunan Province confirmed organic fertilizers (both low/high application) outperform inorganic counterparts in boosting MBC (Huang et al. 2012). However, some research reports insignificant effects on bacterial abundance, suggesting outcomes depend on soil type and fertilizer composition. Long-term organic fertilization may significantly alter bacterial community structure while reducing diversity (Wang et al. 2011; Zhang et al. 2010).

1.3.2.3 Impacts of Other Soil Conditioners

Soil conditioners profoundly influence SOM and microbial community structure. Since microorganisms mediate SOM stabilization/utilization, conditioner-induced soil remediation largely relies on microbial communities. Thus, investigating how conditioners alter microbial structure, abundance, and diversity is essential for deciphering their mechanisms in regulating soil properties and nutrient dynamics.

Conditioners enhance crop yield through strong correlations with microbial richness, significantly promoting microbial proliferation. They also increase functional diversity indices (richness and evenness) in red soils (Jin et al. 2017). Advanced sequencing technologies now enable comprehensive microbial diversity analysis. Metagenomic examination of biochar-amended tobacco soils revealed consistently elevated microbial populations and diversity across growth stages, coupled with reduced bacterial wilt incidence (Shen et al. 2017).

Further research is needed to elucidate how conditioners affect microbial abundance/diversity. Genomic sequencing of soil microbes can fully characterize community structure, composition, and succession, providing analytical tools to decipher conditioner-induced microbe-SOM interactions.

References

Cai X, Yang J, Ma W, Wang H, Xu W (2008) Effects of nitrogen supply levels and water schemes on rice rhizosphere microbial biomass carbon in rice development stage at paddy field. J Zhejiang Univ (Agric Life Sci) 34(6):662–668

Chang Y, Huang H, Zhao Z, Yin J, Zhang H, Zhang J (2021) Prospect for the development of recycling treatment and high-value utilization technologies of food waste. Environ Sanitation Eng 29(1):44–51

Cogger C, Brown S (2016) Soil formation and nutrient cycling. Springer Nature

Deng J (2019) Harmless treatment and resource utilization of kitchen waste: development status and trend. J Environ Eng Technol 9(6):637–642

Hou J, Li M, Mao X, Hao Y, Ding J, Liu D, Xi B, Liu H (2017) Response of microbial community of organic-matter-impoverished arable soil to long-term application of soil conditioner derived from dynamic rapid fermentation of food waste. PLoS ONE 12(4):e0175715

Huang W, Chen A, Wang W, Xie X, Tong C, Liu S, Su Y (2012) Effect of long-term fertilization on active organic carbon and nitrogen in paddy soils. 31(9):1854–1861

Jia W (2008) Effects of long-term fertilization on the change law of the microbial biomass C, N and soil enzyme activity in cinnamon soil. Shanxi University, Shanxi

Jin H, Qi S, Zhu Y, He Q, Liang Z, Dan J (2017) Effects of Soil conditioner on soil nutrient content and microbia functional diversity. J Agric Sci Technol 19(5):100–105

Pan X, Zhang M, Zhang J, Li Y, Jiang F, Yang X, Huang H (2024) Research progress of food waste anaerobic digestion technology. Guangdong Chem Indus 51(5):114–115

Shen J, Lu A, Li X, Li H, Zeng Q, Wang Y (2017) Effect of soil conditioner on the microbial community in the rhizosphere soil of tobacco. Hubei Agric Sci 56(1):39–42

Wang F, Li J, Zhao S, Shen Q, Chu G, Wei C, Guo Y (2011) Effects of antagonistic bacillus and bio-organic fertilizer on controlling cotton *Virticillium Wilt* and activity of soil enzyme. Chin J Soil Sci 42(3):584–588

Wang Y, Ren L, Zhao B (2017) Effect of initial pH and temperature on biohydrogen production from food waste by anaerobic fermentation. Chin J Environ Eng 11(12):6470–6476

Xin Z, Han G, Diao W (2023) Study on resource utilization path of municipal food waste. Sichuan Environ 42(4):344–350

Zhang Q, Wang X, Shi Y, Wang G (2010) Effects of different fertilizer treatments on ecological characteristics of microorganism in chemical fertilizer omission paddy soil. Plant Nutr Fertilizer Sci 16(1):118–123

Zhao J, Yu Z, Li Y, Wang X (2006) Effects of nitrogen application rate on soil inorganic nitrogen distribution, microbial biomass nitrogen content and yield of wheat. Plant Nutr Fertilizer Sci 12(4):466–472

Chapter 2
Organic Components and Microenvironment During Aerobic Fermentation of Food Waste

Abstract Aerobic composting is a process in which microorganisms degrade organic matter in the raw materials under aerobic conditions to produce humus. This process not only eliminates pathogens present in the feedstock but also removes odorous gases and other harmful substances, serving as an important approach for the reduction, harmlessness, and resource recovery of organic waste. This chapter introduces three typical aerobic composting technologies: traditional aerobic composting, inoculant-enhanced composting, and rapid thermophilic composting. It analyzes the classification and transformation of organic carbon components during aerobic composting, with emphasis on the pathways of organic carbon transformation and the succession patterns of microbial communities in food waste under different aerobic composting processes.

2.1 Typical Aerobic Fermentation Processes for Food Waste

2.1.1 Conventional Aerobic Fermentation

Aerobic composting is an important pathway to simultaneously achieve the reduction, valorization, and harmless treatment of organic solid waste. Through aerobic composting, organic solid waste can be processed to produce natural humus for conditioning arable soil. The aerobic composting process involves the microbial degradation of organic matter in the compost material within an oxygen-rich environment. This process can also eliminate pathogens in the compost pile and has a removal effect on odorous gases and harmful substances in the organic solid waste.

The main influencing factors in the aerobic composting process are temperature, moisture content, C/N ratio, and pH. Moisture content is a critical factor; generally, the aerobic composting process can be achieved when the moisture content is between 50 and 70%. A suitable C/N ratio is also a prerequisite for aerobic composting. Microorganisms require sufficient carbon and nitrogen sources for survival and growth, and maintaining a C/N ratio between 20 and 35 helps facilitate the process. Temperature also has a significant impact; an initial temperature that is too low is

J. Hou et al., *Food Waste Valorization: Preparation, Properties, and Application of the Synergistic Soil Conditioners*, https://doi.org/10.1007/978-981-95-5362-4_2

not conducive to microbial survival and makes it difficult to complete the microbial start-up process. The composting process is generally divided into four stages based on temperature: the warming phase, the thermophilic (high-temperature) phase, the cooling phase, and the maturation phase. The total composting time is 45–60 days. The cooling and maturation phases are relatively long, and the main microbial reaction that occurs during these stages is the degradation of recalcitrant materials such as lignin.

Aerobic composting methods mainly include natural composting, windrow composting, agitated-bay composting, and in-vessel composting. Natural composting is the most traditional method, where organic solid waste is piled in a flat, dry area to ferment and decompose under natural conditions. This method is easily affected by weather, temperature, and other environmental factors, and the maturation period is long. Windrow composting is an improvement on the traditional method. Organic solid waste is piled into trapezoidal windrows, spaced at certain intervals. The bottom is layered with breathable organic material to meet ventilation requirements. It is simple to operate and low-cost, but it requires a large area and has a long maturation period. Agitated-bay composting involves mixing the organic solid waste and placing it into a fermentation trench, where the material is turned or aerated regularly to achieve maturation. This technology is highly mechanized and makes it easy to control fermentation conditions like temperature and moisture content, but the operation is complex and it requires a large area. In-vessel composting is currently a more advanced method. Organic waste is placed in a specific reactor for aerobic fermentation. The reactor is generally equipped with sensors for temperature, humidity, etc., allowing for precise control of reaction conditions and effective prevention of secondary pollution such as odorous gases. It is highly automated; however, in-vessel composting technology is not yet mature. Firstly, the limited volume of the reactor leads to low treatment efficiency. Secondly, the manufacturing cost and operational energy consumption of the reactor are high, so large-scale application is currently limited. Nevertheless, the technology's characteristics of precise reaction and targeted fermentation give it broad application prospects.

2.1.2 Inoculation-Enhanced Fermentation

To address the slow start-up time and long processing cycle of traditional fermentation methods, exogenous microbial inoculants are often introduced to enhance the humification efficiency. Currently, the microbial agents inoculated during the humification of organic waste mainly fall into two categories: one is the expanded culture of single or multiple effective microorganisms, and the other consists of products from a completed fermentation and maturation process.

The addition of exogenous inoculants affects humification efficiency mainly in the following ways: (1) It rapidly stimulates functional microbial communities and extends the duration of the thermophilic fermentation phase. Under the action of indigenous microorganisms, the degradation of organic matter causes the temperature of the fermentation material to rise. Inoculating with exogenous microorganisms increases the heating rate to some extent and prolongs the duration of the high-temperature period. (2) It accelerates the humification process and shortens the fermentation cycle (Xi et al. 2003). Currently, single pure cultures of strains produce fewer cellulose-degrading enzymes, resulting in unsatisfactory degradation performance. When multiple cellulose-degrading bacteria are cultured together, a variety of enzymes are produced. Through their synergistic action, the conversion efficiency of cellulosic materials is significantly improved. Inoculating microbial agents at different reaction stages can accelerate cellulose decomposition and speed up the humification process. For example, Li inoculated a self-made exogenous microbial agent during chicken manure fermentation, which accelerated the heating rate, extended the high-temperature fermentation time, and ultimately enhanced the humification process (Li 2014). Yue et al. added a high-efficiency composite cellulose-degrading microbial agent to compost material, which significantly increased the pile temperature and shortened the fermentation time while meeting harmlessness standards (Yue et al. 2018). Inoculating with exogenous strains can promote humification efficiency to a certain extent. However, the sources of materials for humification are diverse and structurally complex. The inoculated exogenous microorganisms have environmental adaptability issues, and there is a certain stimulation effect (including synergistic and inhibitory effects) between them and the indigenous microorganisms (Xi et al. 2003; Song et al. 2016). Therefore, we propose that a multi-stage inoculation method can prolong the high-temperature period and effectively avoid competition between the inoculated agents and indigenous microorganisms, which is conducive to the growth of dominant microorganisms (Xi et al. 2015). Moreover, through the synergistic action of multiple microorganisms, the exogenous inoculant can effectively alleviate problems such as acidification inhibition and low humification efficiency.

2.1.3 Rapid Hydrothermal Composting

Rapid Hydrothermal Composting (RC) is a technology that rapidly degrades organic solid waste (such as food waste, agricultural waste, and sludge) through microbial metabolism under high-temperature (typically 50–80 °C), high-humidity (moisture content 50–70%), and aerobic conditions. Its core principles include: Hydrothermal environment accelerates hydrolysis: High-temperature and high-humidity conditions promote the hydrolysis of organic matter (such as cellulose, protein, and fat) into small-molecule substances (such as monosaccharides, amino acids, and fatty acids); Efficient microbial metabolism: Thermophilic bacteria (such as Bacillus and Actinomycetes) exhibit enhanced activity at high temperatures, rapidly decomposing organic matter and shortening the fermentation cycle; Energy and material cycling: The products can be organic fertilizer, biogas, or high-value-added chemicals (such as volatile fatty acids and lactic acid).

Rapid hydrothermal composting is typically divided into the following stages:

Pre-treatment: Crushing and screening to remove impurities, and adjusting the moisture content to around 60%.
Heating and inoculation: Heating to the target temperature (e.g., 60 °C) and inoculating with a high-efficiency composite microbial agent (e.g., cellulose-decomposing bacteria, thermophilic bacteria).
Aerobic fermentation: Aerating the pile, maintaining the temperature at 55–70 °C for a cycle of 3–7 days. The product is a humic acid-type organic fertilizer.
Post-treatment: Solid–liquid separation, deodorization, drying, or further valorization.

A food waste treatment project in Beijing, China, uses a "hydrothermal fermentation + aerobic composting" process to treat 200 tons per day, with a 7-day fermentation cycle, producing organic fertilizer for landscaping. Shenzhen has piloted a "hydrothermal anaerobic digestion" technology, which has increased the efficiency of biogas production from kitchen waste by 30%. The "Bio-eco" system in Yokohama, Japan, uses 70 °C hydrothermal fermentation to treat raw garbage, achieving a 50% volume reduction within 24 h, with the residue used as a soil conditioner.

Rapid hydrothermal composting technology achieves efficient reduction and valorization of organic solid waste through the synergistic action of a high-temperature, high-humidity environment and microorganisms. It is particularly suitable for treating easily perishable organic matter such as food waste and agricultural waste. Despite challenges like energy consumption and microbial agent compatibility, its advantages of a short cycle and high-value products make it an important development direction in the field of organic solid waste treatment. Future efforts should combine policy support with technological innovation to further promote its large-scale application.

2.2 Classification and Transformation of Organic Carbon Components During Different Treatment Processes

2.2.1 Classification of Organic Carbon Components

Organic carbon components are typically classified using four primary methods: (1) based on compound polarity and charge characteristics; (2) based on molecular weight (MW); (3) categorized as Labile Organic Carbon (LOC) and Recalcitrant Organic Carbon (ROC); and (4) based on component density.

(1) Classification Based on Polarity and Charge Characteristics

This method employs anion-exchange resins, cation-exchange resins, and XAD resins for separation. Based on the charge properties (acidic/basic) and hydrophobicity/hydrophilicity of the organic matter, it can be divided into six fractions:

hydrophobic acids, hydrophobic bases, hydrophobic neutrals, hydrophilic acids, hydrophilic bases, and hydrophilic neutrals.

(2) Classification Based on Molecular Weight (MW)

Molecular weight is a key physicochemical property of organic carbon components, often closely correlated with their function.

1. Macromolecular Organic Carbon: Primarily refers to organic carbon with MW > 5000 Da, including complex organic substances such as humic acids and fulvic acids. This fraction constitutes a significant proportion in mature composts and is largely composed of inert humic-like organic components (e.g., humic acid substances). Their high molecular weight arises from the condensation of aromatic groups, resulting in a significantly higher degree of aromatization compared to components with MW < 5000 Da. This imparts structural stability and renders them resistant to direct microbial utilization. An MW > 5000 Da is often regarded as a demarcation standard for the maturity of fermentation products; specifically, a higher content of this fraction indicates a greater degree of humification. It provides slow-release nutrients that promote soil aggregate formation. During fermentation, the decomposition of protein-like substances occurs synergistically with the aromatization of humic-like substances.
2. Medium Molecular Weight Organic Carbon: Comprises organic carbon with MW between 600 and 5000 Da, mainly including amino acids, fatty acids, monosaccharides, and oligosaccharides.
3. Low Molecular Weight Organic Carbon: Refers to organic carbon with MW < 600 Da. These components can be directly absorbed and utilized by microorganisms through passive or active transport across cell walls, or they can directly participate in chemical transformation processes, playing a vital role in nutrient cycling. Macromolecular complexes such as carbohydrates, proteins, nucleic acids, and lipids derived from plant residues or marine phytoplankton require enzymatic degradation into substrates like monosaccharides, oligosaccharides, or amino acids with MW < 600 Da to become utilizable nutritional substrates for microorganisms.
4. Molecular Weight Dynamic Transformation: Under different process conditions, organic carbon with MW > 5000 Da and MW < 600 Da can undergo interconversion through degradation and polycondensation reactions. During the humification of organic waste, a portion of the organic carbon undergoes direct mineralization, releasing greenhouse gases such as CO_2 and CH_4; Low molecular weight Labile Organic Carbon (LOC) serves as the primary source of energy and nutrients for microbial assimilation (its content is high in the initial humification stage but decreases significantly in intermediate and final stages); Recalcitrant organic carbon (e.g., lignin, cellulose), under microbial metabolism, yields decomposition products that form humic precursor substances, ultimately synthesizing humus (leading to an increase in the content of macromolecular recalcitrant

organic carbon). Consequently, the humification process involves dynamic changes and morphological transformations among organic carbon fractions of varying molecular weights.

(3) Classification Based on Availability: Labile Organic Carbon (LOC) and Recalcitrant Organic Carbon (ROC)

1. Labile Organic Carbon (LOC) primarily includes:

   Dissolved Organic Carbon (DOC): Possesses certain solubility, exhibits high activity, serves as the main carbon source for microorganisms, and represents a relatively dynamic component of soil organic carbon.

   Microbial Biomass Carbon (MBC): Refers to the carbon contained within soil microorganisms (typically representing cells with a volume $< 5 \times 10^3$ μm^3). It is a key indicator reflecting soil carbon stability and nutrient dynamics.

   Easily Oxidizable Organic Carbon (EOC): Defined as the readily oxidizable fraction of organic carbon, primarily composed of low molecular weight compounds derived from plant and microbial activity. It is commonly quantified using oxidation with 0.33 mol L^{-1} $KMnO_4$ or 0.8–1.0 mol L^{-1} $K_2Cr_2O_7$.

   Particulate Organic Carbon (POC): Refers to organic carbon within the particle size range of 0.053–2 mm. Due to its ease of measurement, POC has become a widely used indicator for estimating the labile soil organic matter content.
2. Recalcitrant/Inert Organic Carbon (ROC): Primarily refers to organic carbon components with complex structures that are resistant to rapid decomposition and utilization by microorganisms, such as humic substances.

(4) Classification Based on Density: Light Fraction Organic Carbon (LFOC) and Heavy Fraction Organic Carbon (HFOC)

1. Light Fraction Organic Carbon (LFOC): Has a density lower than 1.8–2.0 g cm^{-3}. It is mainly composed of lignin monomers, lipids, aromatic compounds, etc., originating from free humic acids and incompletely decomposed plant and animal residues. LFOC represents a relatively active organic carbon fraction.
2. Heavy Fraction Organic Carbon (HFOC): Has a density higher than 1.8–2.0 g cm^{-3}. It constitutes an assemblage of organic carbon bound within soil aggregates and mineral particles. Its complex structure makes it difficult for microorganisms to transform or degrade, classifying it as protected organic carbon. HFOC is a crucial indicator for assessing soil carbon sequestration capacity.

Separation and Transformation: These two fractions can be separated using a solution of specific relative density (e.g., NaI heavy liquid): the upper suspension, after drying, yields LFOC, while the lower precipitate constitutes HFOC. The application of exogenous organic amendments to soil influences microbial abundance and activity, increases total soil organic carbon content, and alters the dynamic transformation between LFOC and HFOC components.

2.2.2 *Transformation of Organic Carbon Components During Different Treatment Processes*

Researchers have employed various biological treatment technologies to enhance the fermentation efficiency of organic waste, which significantly alters the content and distribution of organic carbon components in the resulting products.

(1) Microbial Inoculation Technology

Microbial inoculants enhance organic matter decomposition during fermentation, promote the generation of low-molecular-weight organic carbon (LMWOC), accelerate the synthesis of humic-like substances, and thereby shorten the fermentation cycle.

During swine manure composting, microbial inoculation accelerates the conversion of macromolecular organic matter into low-molecular-weight forms, evidenced by a continuous decrease in macromolecular content and a significant increase in the proportion of small molecules. Inoculating chicken manure compost increased the content of humic acid and fulvic acid, while also elevated the molecular weight, polymerization degree, and aromaticity of the products. EM inoculants could increase humic acid, soluble organic carbon (SOC), and microbial biomass carbon (MBC), alongside a decrease in total carbon (TC) and easily oxidizable organic carbon (EOC) content, and an increase in the proportion of humic acid carbon. An interconversion relationship existed between MBC and water-soluble carbon (WSC).

In summary, microbial inoculation not only regulates the transformation pathways of organic carbon components but also exhibits significant correlations with the aromaticity of the products and the progression of humification.

(2) Earthworm Bio-stabilization Technology

This technology utilizes the synergistic system of earthworm gut microbiota to process organic waste (e.g., food waste). Digestion achieves waste reduction, and the resulting vermicompost serves as a high-quality organic fertilizer. Its core mechanisms involve: Degradation of macromolecular organic matter into bioavailable components and polymerization of small molecular monomers into humus.

Compared to conventional composting, vermicomposting has two key distinguishing features: The transformation process occurs within the in vivo environment of the earthworm gut; a dynamic continuous feeding-discharge mode was operated. When vermicomposting of sewage sludge, an increase in dissolved organic matter (DOM) components < 300 Da and a decrease in components > 8000 Da were both observed. Moreover, an increase in hydrophobic and aromatic substances and a decrease in protein components were also observed. Treatment of livestock manure resulted in a significant reduction in microbial biomass carbon (MBC). Reductions in lipid, protein, and polysaccharide components; enhanced aromaticity and content of oxygen-containing functional groups (e.g., carboxyl, phenolic hydroxyl, ether linkages) within the humus, promoting the formation of stable humic-like structures.

The limitation is that earthworms exhibit low reproduction rates and weak environmental adaptability (particularly poor heat tolerance; temperatures > 40 °C can be lethal).

(3) Maggot Bioconversion Technology

Maggots (e.g., the black soldier fly, Hermetia illucens) transform organic matter through a tripartite mechanism: Chewing, turning, and bio-ventilation for physical agitation; secretion of enzymes for carbohydrates, fats, and proteins degradation; and function of fat accumulation. The products have low odor and a loose structure. Following digestion by black soldier fly larvae (BSFL), simple-structured organic components in the frass decrease, while aromaticity and molecular weight increase. BSFL utilize proteinaceous materials and soluble microbial byproducts as substrates to form humic matrices. When treating livestock manure, BSFL improve the degradation of total nutrients and organic matter, reduce volatile fatty acid (VFA) accumulation, and enhance product quality. However, high moisture content and cellulose levels in organic waste significantly impede bioconversion rates.

Based on a comprehensive comparison, microbial inoculation technology demonstrates superior potential for practical implementation in enhancing fermentation efficiency.

2.3 Transformation of Organic Carbon Components During Different Aerobic Fermentation Processes of Food Waste

2.3.1 *Changes in Organic Matter During Fermentation*

Different bioprocessing methods affect not only the dynamic changes in organic carbon components but also the humification process. Using food waste as the raw material and rice husk as the bulking agent at a mass ratio of 5:2, three treatments were conducted: conventional fermentation (CK), fermentation with phased inoculation (JJ), and rapid hydrothermal fermentation treatment (RC). The physicochemical properties of the fermentation materials are shown in Table 2.1.

Table 2.2 outlines the fermentation process and sampling times for the three methods. For all groups, food waste and rice husk were mixed uniformly at a 5:2 ratio before being added to the fermentation units. The CK group was not inoculated

Table 2.1 Physicochemical properties of the fermentation materials

Material	C/N	Organic matter/wt%	Moisture content/%
Food waste	23.5 ± 0.45	87.7 ± 1.23	83.2 ± 1.22
Rice husk	32.4 ± 0.34	89.4 ± 1.34	10.5 ± 0.37
Fermentation feedstock	26.0 ± 0.42	88.2 ± 1.15	62.4 ± 1.04

and underwent natural fermentation for 30 d; the JJ group underwent fermentation with phased inoculation for 15 d, with a 2% acid-resistant bacterial strain inoculated at the initial stage, followed by a 2% cellulose-degrading bacterial strain once the material temperature reached 50 °C; the RC group was inoculated at the beginning of fermentation and underwent rapid hydrothermal fermentation for 24 h.

The organic matter in the materials provides nutrients for the metabolic activities of microbial communities and is mineralized to form CO_2. Therefore, the content of organic matter shows a decreasing trend during fermentation. As shown in Fig. 2.1, the organic matter content in all three fermentation methods showed a fluctuating decrease. In the JJ group, the organic matter content peaked at 87.77% on day 6 of fermentation and decreased to 81% by the end. The reason for this is that during the initial stage of oxidation, water-soluble organic matter accumulates within the material, causing a slight increase in organic matter content. As fermentation proceeds, some organic matter is decomposed and utilized by microorganisms, while another portion is mineralized, leading to CO_2 loss and a reduction in the total amount. The change in organic matter in the RC group was not significant, possibly because the organic components in the rice husk were minimally degraded, and the degradation of organic matter from food waste was insufficient to cause a noticeable change in the overall organic matter of the material.

Table 2.2 Experimental design and sampling methods

Fermentation method	Fermentation process	Sampling times
CK	Fermentation without inoculation for 30 d	1 d, 2 d, 4 d, 6 d, 8 d, 11 d, 15 d, 20 d, 24 d, 30 d
JJ	Fermentation with phased inoculation for 15 d, with an initial inoculation of a 2% acid-resistant bacterial strain, followed by a 2% cellulose-degrading bacterial strain when the material temperature reached 50 °C	1 d, 2 d, 4 d, 6 d, 8 d, 9 d, 11 d, 15 d
RC	Initial inoculation, rapid hydrothermal fermentation for 24 h	1 h, 3 h, 5 h, 7 h, 11 h, 13 h, 24 h

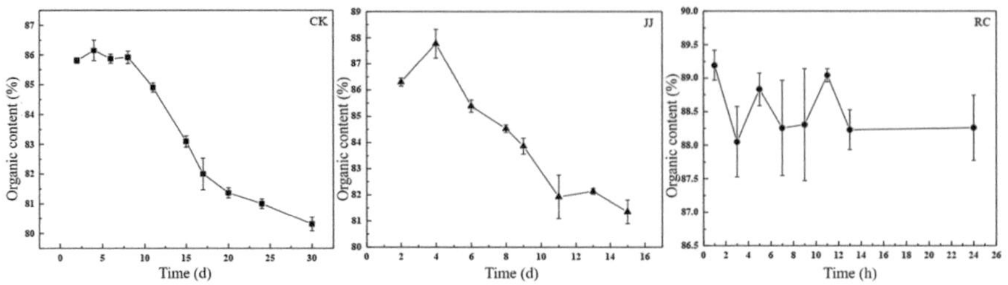

Fig. 2.1 Changes in organic matter during different fermentation processes

2.3.2 Spectroscopic Characterization of Different Organic Carbon Fractions During Fermentation

2.3.2.1 Three-Dimensional Fluorescence Spectroscopy

Fluorescence Excitation-Emission Matrix (EEM) spectroscopy is a rapid, sensitive, and reliable analytical method that does not require complex sample pretreatment. The three-dimensional fluorescence spectra can be categorized into five distinct regions: I, II, III, IV, and V, corresponding to tyrosine-like substances, tryptophan-like substances, fulvic acid-like substances, soluble microbial products (SMPs), and humic acid-like substances, respectively. The EEM spectra reflect the transformation of organic carbon components during the fermentation process, with the $P_{i,n}$ value for each region indicating the relative content of these organic carbon components. Regions I and II are generally associated with protein-like fluorescence peaks, whereas Regions III and V represent humic-like substances. In the early stages of fermentation (Fig. 2.2), low material temperatures suppress microbial activity, resulting in slow protein and amino acid decomposition and the formation of only a small amount of humic-like substances. Therefore, tyrosine and tryptophan substances in Regions I and II are the primary fluorescence peaks of CK. As fermentation progresses, the fluorescence peak intensities in Regions I, II, and IV gradually decline, whereas the fluorescence peaks of humic-like substances in Regions III and V gradually increase, eventually becoming the primary fluorescence peaks. This indicates the presence of a large amount of humic-like substances at the end of fermentation.

As shown in Fig. 2.3, the fluorescence intensity of each component was analyzed using Fluorescence Regional Integration (FRI), with the percentage denoted as $P_{i,n}$. At the beginning of fermentation, the fluorescence intensity volume percentage of $P_{I+II,n}$ protein-like substances was relatively high at 30.44%. As microorganisms decomposed them, their content gradually decreased to 5.83% by the end of fermentation. Specifically, tyrosine in region $P_{I,n}$ decreased from an initial 18.10% to 1.82%, and tryptophan substances in region $P_{II,n}$ decreased from an initial 12.34% to 4.01%. The decreasing trend of $P_{I,n}$ was more pronounced than that of $P_{II,n}$, with a significant drop from 16.34% to 5.03% during the high-temperature phase. The reason for this is that tyrosine, as a small-molecule organic substance, can be preferentially utilized by microorganisms, and the degradation rate is faster due to stronger microbial activity during the high-temperature stage. The decreasing trend of $P_{II,n}$ was not significant during this phase, with the content only reducing by 5.88%, possibly because tyrosine-like substances are more easily transformed by microorganisms than tryptophan-like substances, leading to a more significant decrease in the former's content. The content of soluble microbial byproducts, represented by region $P_{IV,n}$, decreased significantly by 20.10%. Its change trend was similar to that of protein-like substances, with a more pronounced decrease during the high-temperature phase and a gentler decline during the cooling and maturation phase of fermentation. In contrast, the content of humic-like acid in region $P_{V,n}$ increased significantly as fermentation proceeded, rising from an initial 26.25% to 66.36%, an increase of 40.11%, making it the most abundant component. In comparison, the content of fulvic-like acid substances in region $P_{III,n}$ did not increase substantially, rising from an initial 11.21–15.81%. This indicates that as fermentation progresses, organic carbon components are gradually transformed into humic-like substances with larger molecular weights and more complex structures, which is the main transformation trend for organic carbon components during fermentation, accompanied by the formation of fulvic-like acid substances. A study has found that humic-like

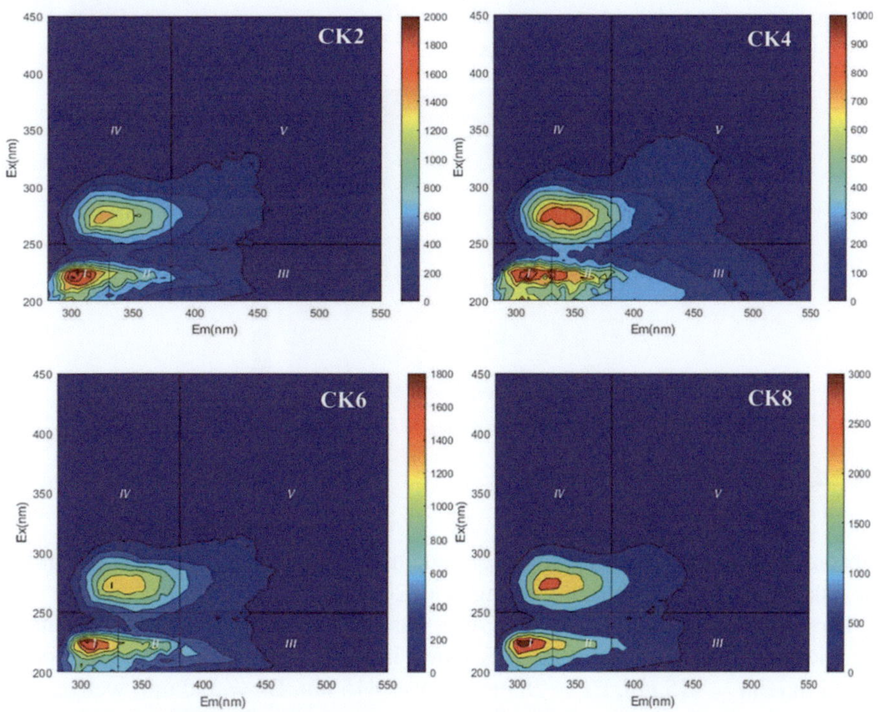

Fig. 2.2 3D fluorescence spectra of organic carbon components for CK

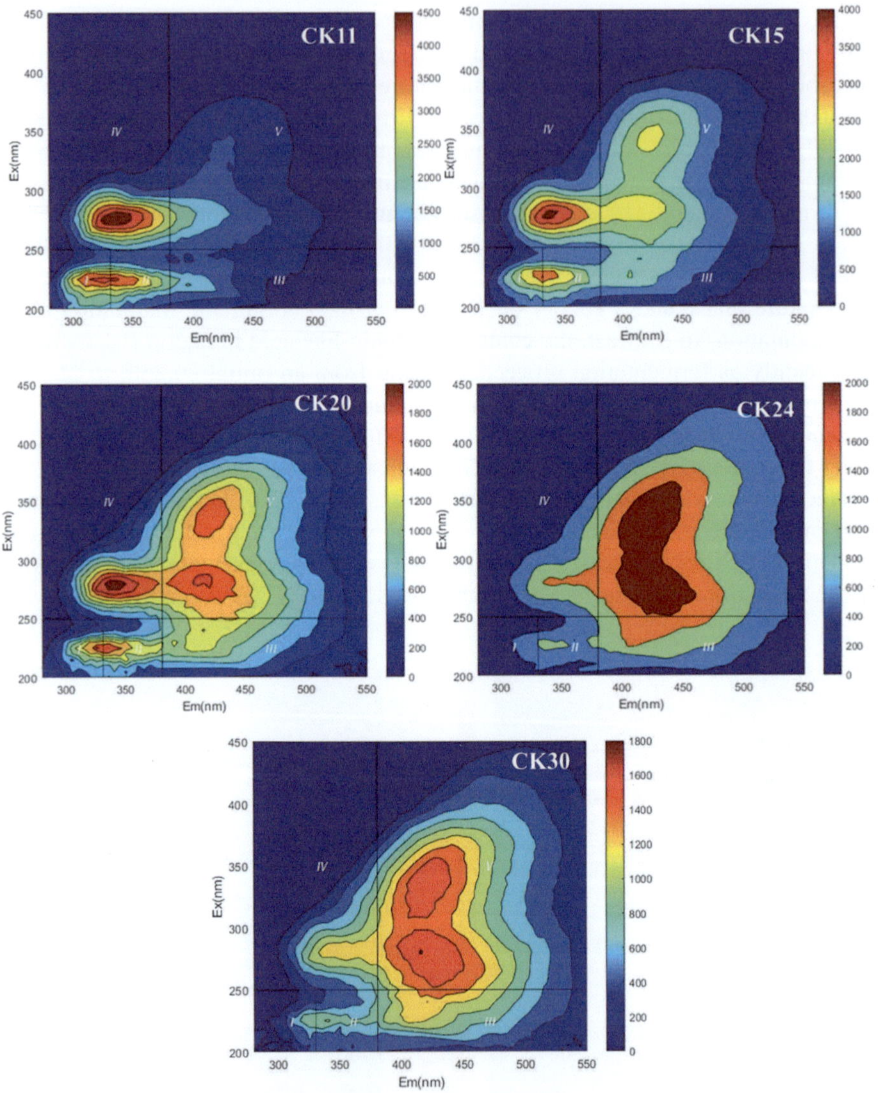

Fig. 2.2 (continued)

substances can not only improve the organic matter content of soil but also regulate the balance of mineral nutrients (Kong et al. 2019).

Parallel Factor Analysis (PARAFAC) was used to further analyze the fate of organic carbon components in CK. As shown in Fig. 2.4, the two fluorescence peaks for C1 at (Em = 415 nm, Ex = 285 nm; Em = 415 nm, Ex = 330 nm) were caused by fulvic-like acid substances. The fluorescence peaks for C2 at (Em = 335 nm, Ex = 225 nm; Em = 335 nm, Ex = 275 nm) were contributions from soluble microbial

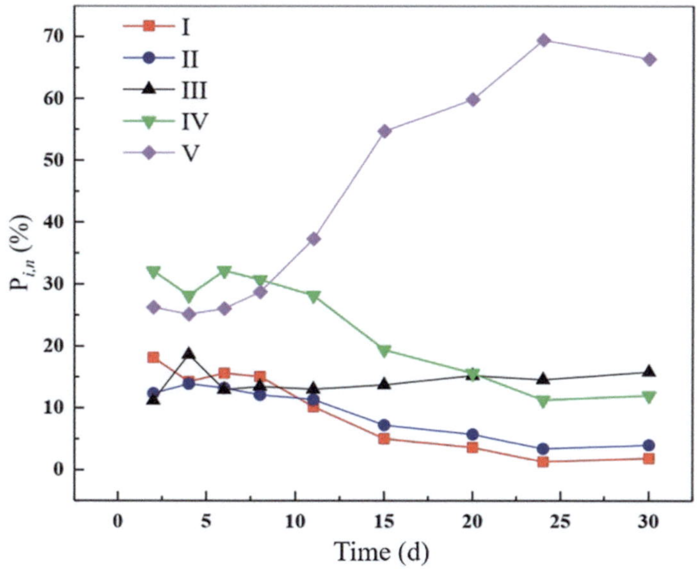

Fig. 2.3 Percentage of fluorescence intensity from regional volume integration for CK organic carbon components

byproducts and tryptophan components, respectively. The fluorescence peaks for C3 at (Em = 455 nm, Ex = 265 nm; Em = 455 nm, Ex = 370 nm) represented humic-like components. The peak for C4 at (Em = 300 nm, Ex = 225 nm) was the fluorescence peak of protein-like substance components.

Figure 2.5 shows that as fermentation proceeded, component C2 (soluble microbial byproducts and tryptophan) and C4 (protein-like substances) decreased significantly. The content of the C4 protein-like component dropped remarkably from 53.27% at the beginning of fermentation to 4.28% at the end, while the contents of C1 (fulvic-like acid substances) and C3 (humic-like substances) increased significantly. This indicates that during the fermentation process, soluble microbial byproducts and protein-like substances were decomposed, while at the same time, under the action of microorganisms, some recalcitrant substances polymerized and transformed into humic-like substances, increasing their content, degree of humification, and aromaticity.

Figure 2.6 shows the 3D fluorescence spectra of organic carbon component changes during the fermentation of the JJ treatment group. The figure shows that in the early stage of fermentation, the fluorescence peak intensities in regions I, II, and IV were high, especially the peak for soluble microbial byproducts in region IV, which was the strongest. The reason for this is that the microorganisms inoculated at the initial stage of fermentation rapidly adapted to the microenvironment in the material and decomposed organic matter, producing a large amount of soluble microbial byproducts. As fermentation continued, the fluorescence peaks in regions I, II, and IV gradually weakened, while the peaks in regions III and V gradually strengthened.

This indicates that protein-like substances and soluble microbial byproduct components gradually decreased, while humic-like components gradually increased. By the end of fermentation, the fluorescence peak in region V was the strongest, indicating a high degree of humification.

PARAFAC was used to analyze the transformation patterns of organic carbon components in the JJ group. The PARAFAC method can effectively avoid the overlap between fluorescence peaks, making quantification more scientific. Figure 2.7 shows the evolution of organic matter components during the fermentation process. PARAFAC divided the organic carbon components into C1, with fluorescence peaks at (Em = 335 nm, Ex = 225 nm; Em = 335 nm, Ex = 275 nm), which are contributions from soluble microbial byproducts and tryptophan components, with the tryptophan fluorescence peak being slightly stronger. The two fluorescence peaks for C2 at (Em = 420 nm; Ex = 285, 330 nm) were caused by fulvic-like acid substances. C3 at (Em = 300 nm; Ex = 225 nm, 275 nm) was mainly composed of protein-like substance components, accompanied by humic-like substances. The fluorescence peak for C4 at (Em = 460 nm; Ex = 265 nm, 375 nm) represented humic-like substances.

As shown in Fig. 2.8, as the fermentation process advanced, the content of component C1, soluble microbial byproducts and tryptophan, showed a fluctuating decrease. Its content increased from 45.43% after the inoculation of the microbial agent on

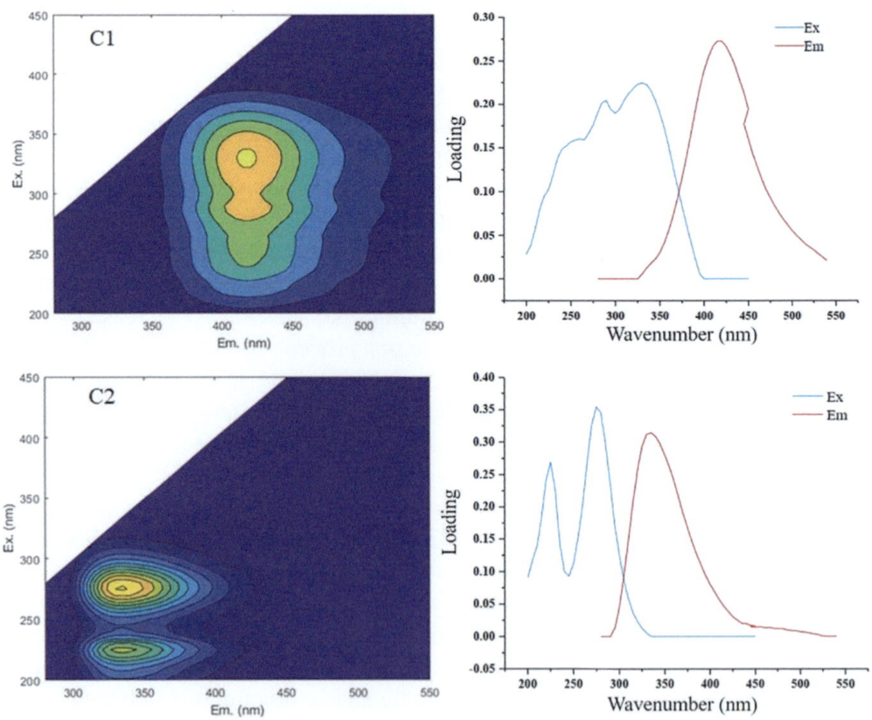

Fig. 2.4 Parallel factor analysis of organic carbon components for CK

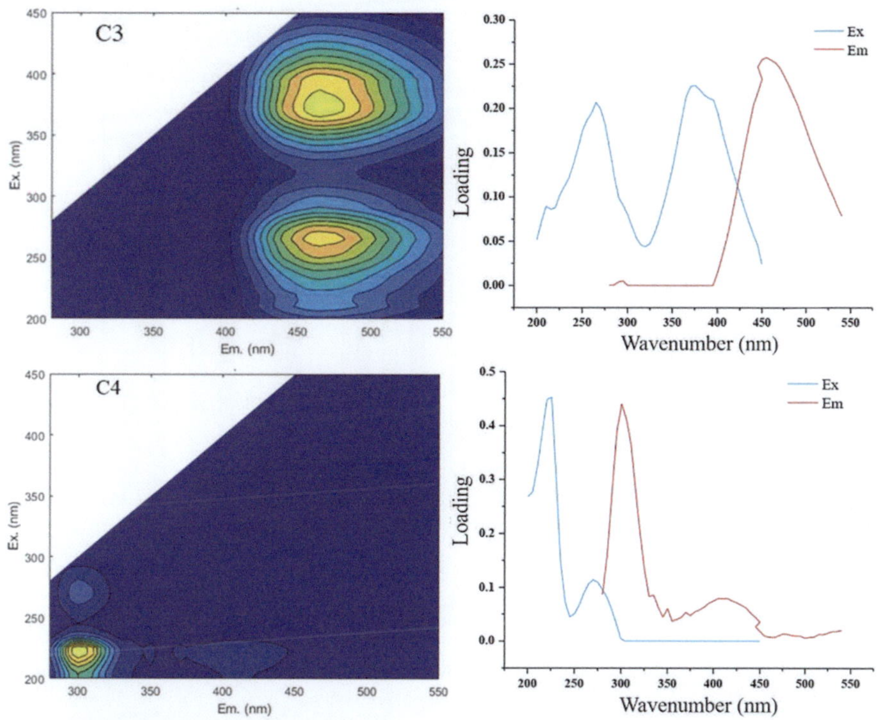

Fig. 2.4 (continued)

day 4 to 57.20% at the beginning of the cooling phase. This indicates that the microbial activity in the material was high, and the microorganisms rapidly adapted to the microenvironment and proliferated, decomposing a large amount of easily degradable organic matter, causing the content of soluble microbial byproducts to reach its peak. This provided a good foundation for the synthesis of humic-like substances in the later stage of fermentation. The content of component C2, fulvic-like acid substances, showed an increasing trend. Especially in the later stage of fermentation when the content of component C1 decreased, its content increased substantially from 15.97 to 35.40%. The fluorescence peak at Ex/Em = 285/420 nm belongs to UV-region fulvic acid, while the peak at Ex/Em = 330/420 nm is produced by visible-region fulvic acid. The UV-region fulvic acid peak originates from small-molecule, high-fluorescence-efficiency substances, while the visible-region fulvic acid peak is a contribution from large-molecule, aromatic fulvic-like acid substances. The peak intensity at Ex/Em = 330/420 nm was slightly greater than that at Ex/Em = 285/420 nm, indicating that the content of high-molecular-weight fulvic acid in the visible region was slightly higher than that of low-molecular-weight fulvic acid substances in the UV region, becoming the main component of the fermentation product. Fulvic-like acid has the characteristics of a small molecular weight, is easily utilized by microorganisms, and has a high content of functional groups (hydroxyl and carboxyl), with phenolic

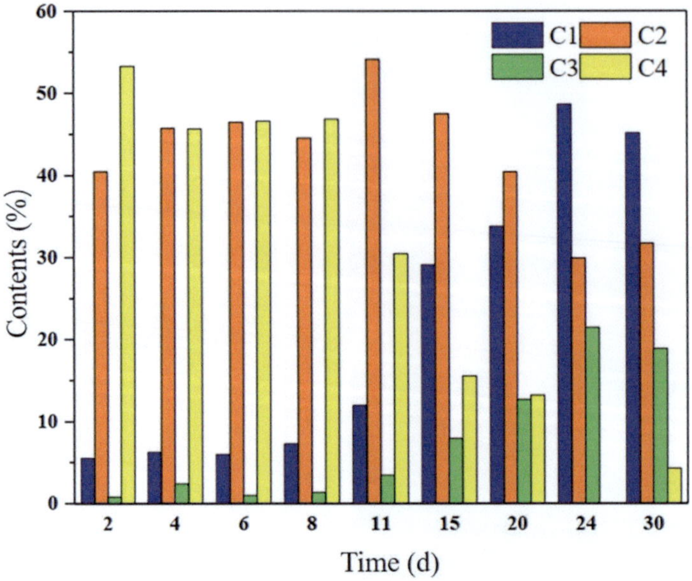

Fig. 2.5 Relative percentage of organic components for CK

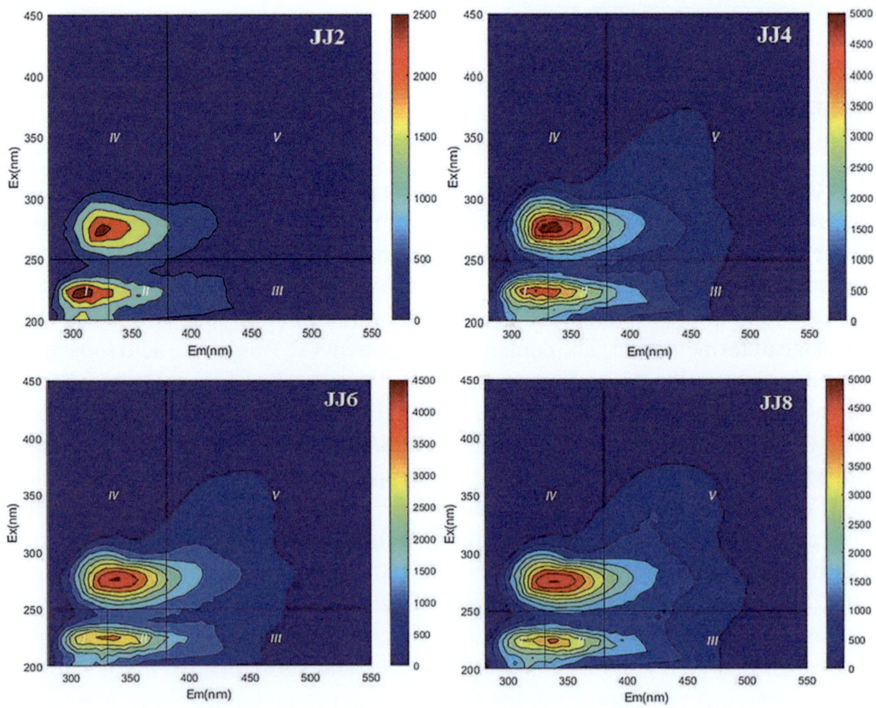

Fig. 2.6 3D fluorescence spectra of organic carbon components for JJ

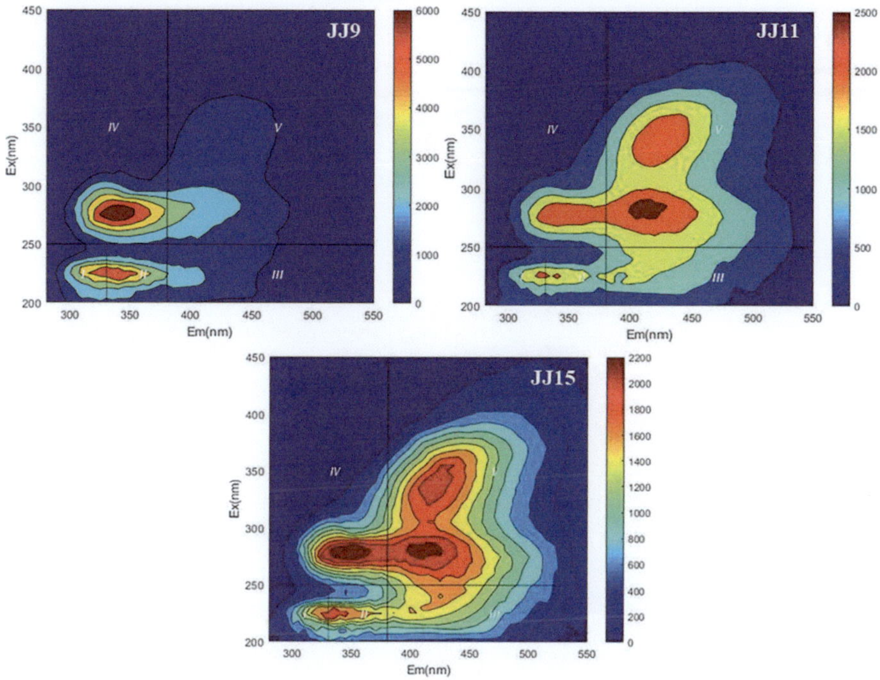

Fig. 2.6 (continued)

hydroxyl groups having growth-promoting functions, leading to better germination effects. This may be one of the main reasons for the higher seed germination rate of JJ. Meanwhile, the content of the C3 component, protein-like substances, decreased significantly, dropping to 12.57% at the end of fermentation, while the content of humic-like substances, C4, gradually increased. This indicates that the characteristics of the fermentation are the decomposition of protein-like substances and the synthesis of humic-like components.

As shown in Fig. 2.9, at the beginning of the RC treatment group's fermentation, the fluorescence peak in region IV was strong, and the content of soluble microbial byproducts was high. As the fermentation process progressed, the intensity of the fluorescence peaks in regions I, II, and IV gradually weakened, the peak in region III showed a slight enhancement, and the peak in region V gradually evolved into the main fluorescence peak. This indicates that during the fermentation process, protein-like substances and soluble microbial byproduct components may have undergone transformation or degradation, causing the fluorescence peak intensity to weaken. The gradual strengthening of the fluorescence peaks in regions III and V may be due to an increase in the content of fulvic-like acid and humic-like acid components.

As can be seen from Fig. 2.10, compared to RC1, regions III and V in RC24 increased by 1.82% and 3.33% respectively, while regions I, II, and IV decreased by

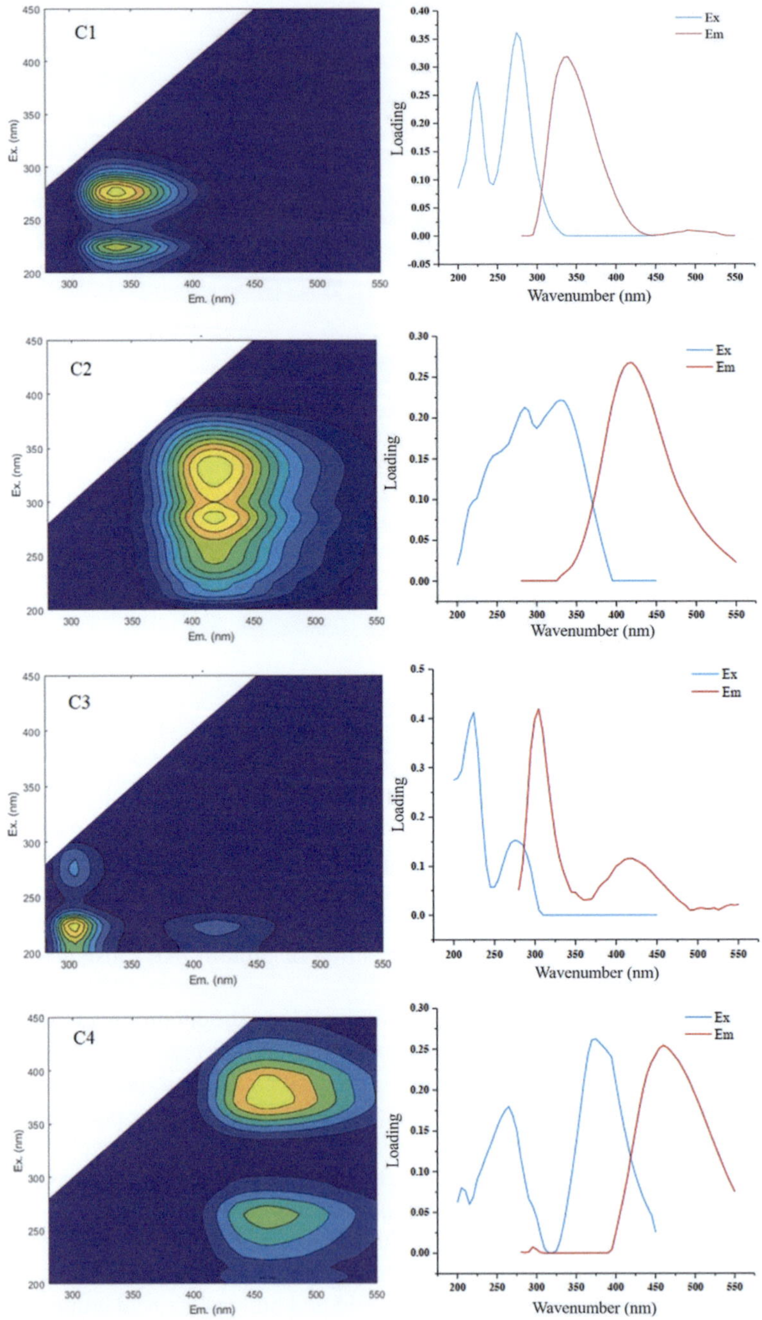

Fig. 2.7 Parallel Factor Analysis of Organic Carbon Components for JJ

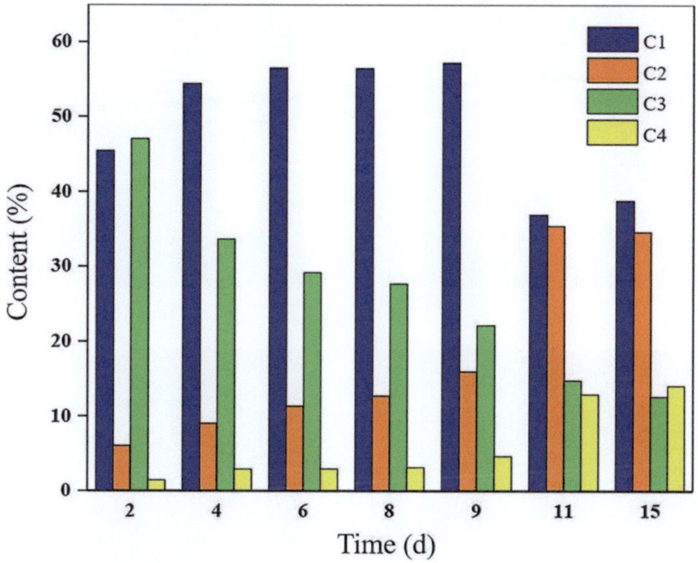

Fig. 2.8 Relative percentage of organic components for JJ

0.42%, 1.88%, and 2.85% respectively. The components I (tyrosine) and II (tryptophan) showed a fluctuating decrease, the soluble microbial byproducts in region IV showed a continuous decreasing trend, and the content of humic-like substances in regions III and V showed an increasing trend. Among them, the content of fulvic-like acid substances in region III reached its highest level at the 5th hour, while at this time, the content of humic-like acid substances in region V dropped to its lowest. This may be because at this point, both the temperature and organic matter content in the material were high, providing a good microenvironment and sufficient nutrients for microbial reproduction, causing them to proliferate. At this time, the fermentation entered the high-temperature phase, and some recalcitrant organic matter such as cellulose and hemicellulose began to degrade, producing small-molecule organic matter and increasing the content of fulvic-like acid substances. As fermentation proceeded, humic-like acid substances gradually increased, while the content of protein-like substances and soluble microbial byproducts decreased, indicating that the synthesis of humic-like substances began in the later stage of fermentation. At the end of fermentation, the content of soluble microbial byproducts in region IV was relatively high, at 27.64%. The reason for this may be that the fermentation time was short, and a large amount of organic matter was decomposed by microorganisms into small-molecule organic compounds that accumulated in the material, with only a part of these small-molecule organic carbon components participating in the synthesis of humic-like substances.

A Parallel Factor Analysis was performed on the RC treatment group. The results in Fig. 2.11 show that the fluorescence peaks for C1 at (Em = 335 nm, Ex = 225 nm; Em = 335 nm, Ex = 275 nm) represent soluble microbial byproducts and tryptophan

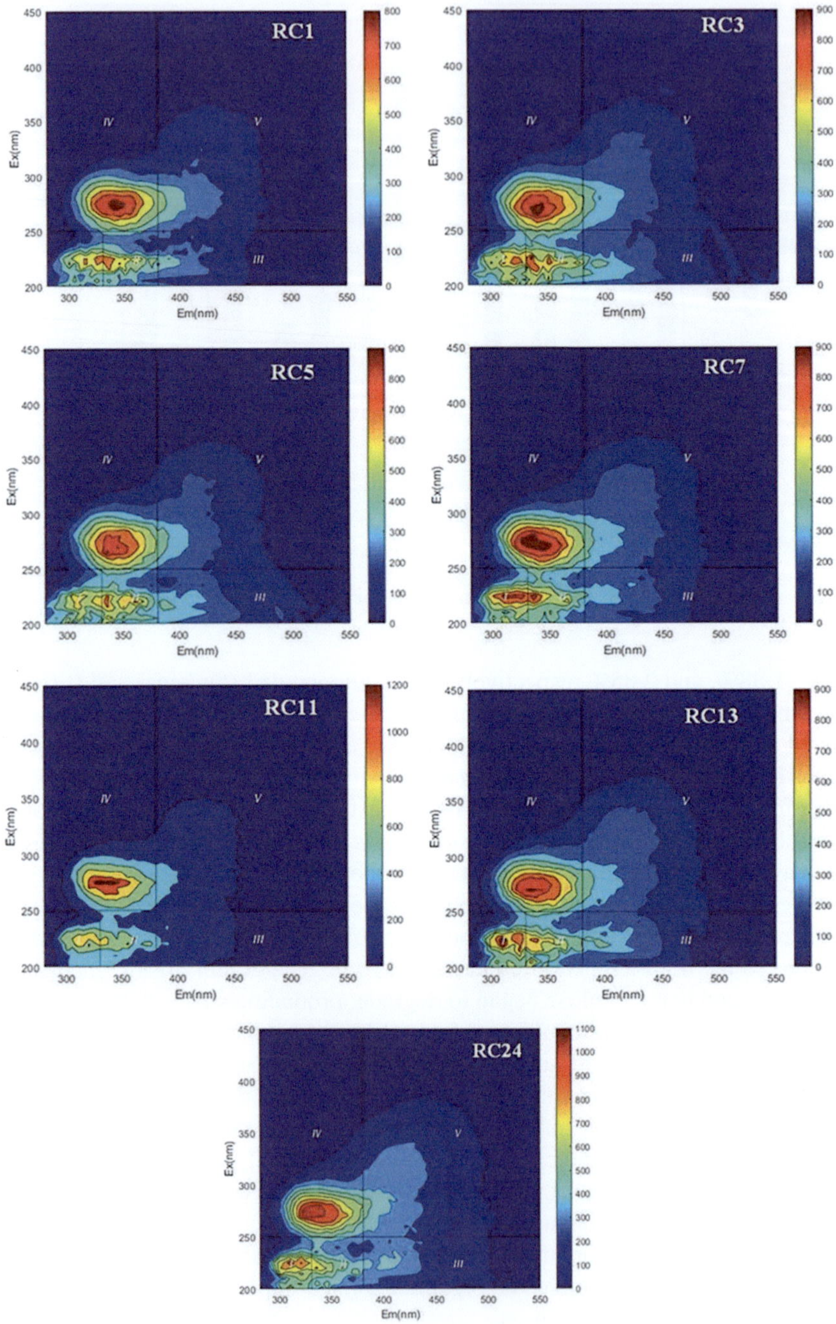

Fig. 2.9 3D fluorescence spectra of organic carbon components for RC

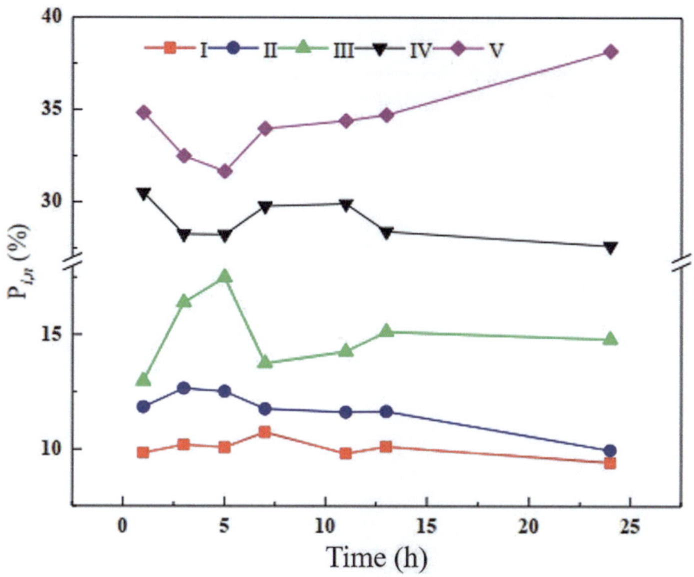

Fig. 2.10 Percentage of fluorescence intensity from regional volume integration for RC organic carbon components

components, respectively. The shoulder peaks for C2 at (Em = 420 nm, Ex = 320 nm; Em = 420 nm, Ex = 245 nm) are contributions from fulvic-like acid and humic-like substances, respectively. The fluorescence peaks for C3 at (Em = 410 nm, Ex = 225 nm; Em = 365 nm, Ex = 280 nm) represent the generation of humic-like acid substances and protein-like substances, with the humic-like acid substance fluorescence peak being stronger.

Figure 2.12 shows the evolution of organic matter components during the fermentation process. Component C1, soluble microbial byproducts and tryptophan, showed a decreasing trend, dropping from an initial 56.50–50.08%. The contents of C2 (fulvic-like acid) and C3 (humic acid) increased, but the change was not significant. Therefore, by the end of fermentation, soluble microbial byproducts and tryptophan substances were the main organic carbon components in the material, which is consistent with the analysis results from the regional volume integration.

Unlike the other two treatment groups, the final fermentation product of RC had a large-molecule humic-like acid content of only 38.17%. Small-molecule organic substances such as protein-like substances and soluble microbial byproducts totaled 47.02%, becoming the main components of the organic matter. This is consistent with the results of molecular weight fractionation. Hou et al. (2017) showed that a large amount of soluble microbial byproducts and carbohydrates can provide readily available nutrients to the soil, promote the growth of microbial communities, and continuously release nutrients, which is beneficial for long-term cultivation. The RC fermentation product had a high content of small-molecule organic carbon, and its degree of humification was much lower than that of CK and JJ. In CK, the content

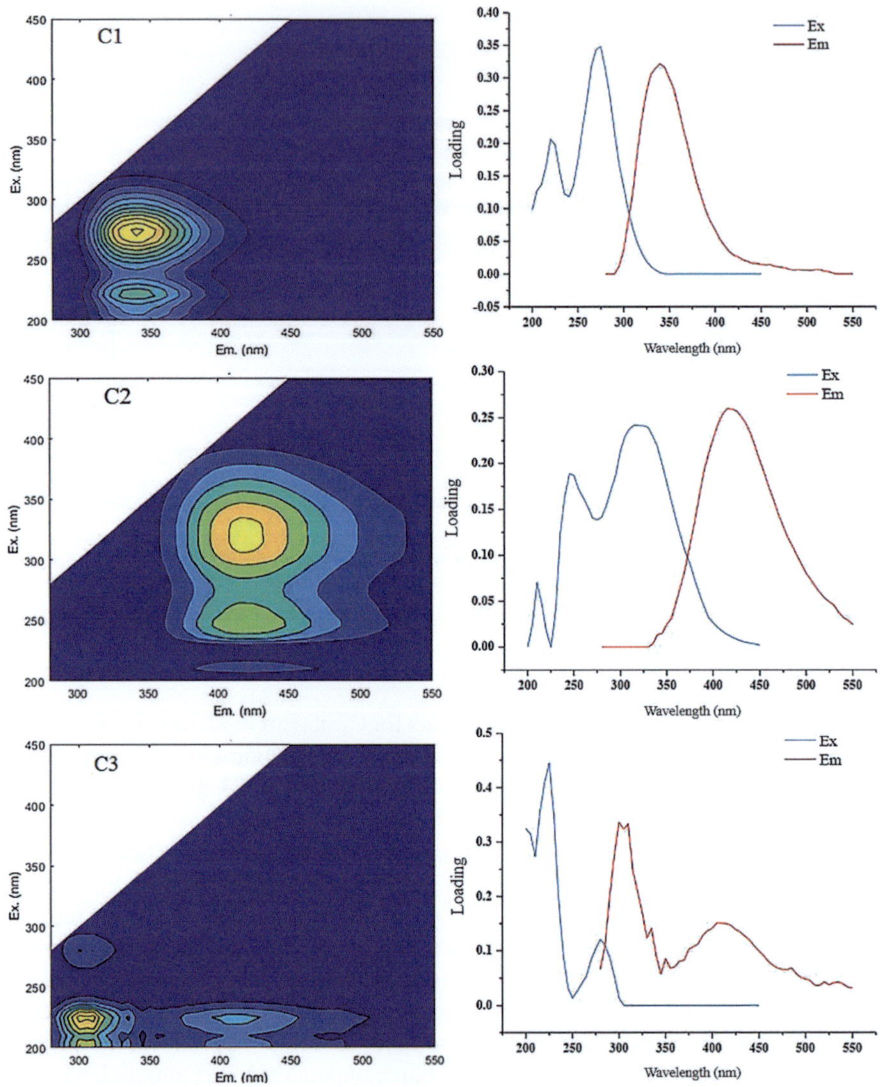

Fig. 2.11 Parallel factor analysis of organic carbon components for RC

of humic-like acid substances was high, accounting for about 66.36% of the total organic carbon components and becoming the main component in the final product. Because its fermentation time was longer, it provided sufficient time for the synthesis of humic substances in the later stage, resulting in a higher degree of humification. In contrast, for the JJ treatment group, the phased inoculation of bacterial agents accelerated the humification speed of the material, resulting in a content of humic-like substances in the fermented product of about 50%, with fulvic-like acid substances

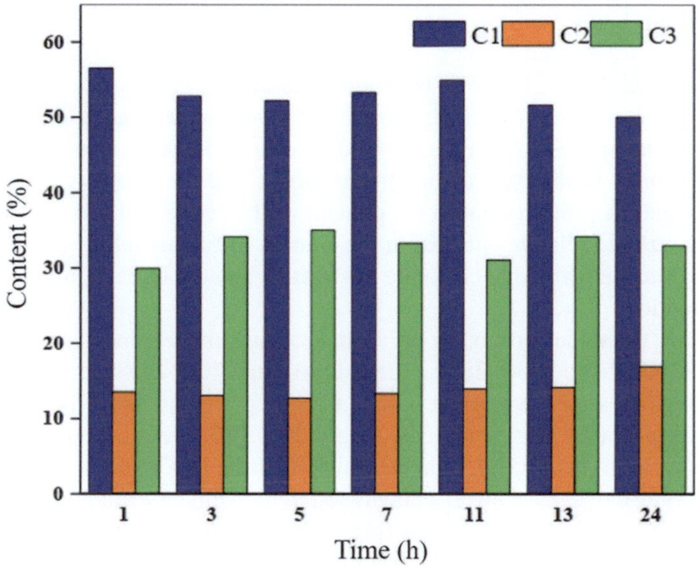

Fig. 2.12 Relative percentage of organic components for RC

being dominant. This indicates that inoculating microbial agents can accelerate the degradation of organic matter, producing a large number of small-molecule organic carbon components. At the same time, due to the shorter fermentation time, some small-molecule organic carbon components polymerize into humic-like substances or form precursors to humic-like substances, which then remain in the material.

2.3.2.2 Fourier-Transform Infrared (FTIR) Spectroscopy

Dissolved organic carbon is a heterogeneous mixture that includes functional groups such as aliphatic, hydroxyl, carboxyl, carbonyl, and ester groups. Fourier-transform infrared spectroscopy (FTIR) is widely used for the characterization of functional groups in organic carbon components due to its ability to perform quantitative and qualitative analysis of major functional groups, such as carboxylic acids, phenols, amides, esters, and saturated/unsaturated hydrocarbons. FTIR was used to further investigate the structural and functional group changes in organic carbon components of different molecular weights. As shown in Fig. 2.13a, c, e, the peak positions for the three treatment groups were similar, with absorption peaks appearing at 3300–3500, 2930–2850, 1650–1630, 1450–1400, 1080–1030, and 870–610 cm^{-1}. The absorption peak for phenolic/alcoholic O–H was in the 3300–3500 cm^{-1} range. The peak at 2930–2850 cm^{-1} was from the C-H stretching vibration of aliphatic methylene groups. The absorption peak for benzene rings, olefinic C=C, or carboxylic acid C=O appeared between 1650 and 1630 cm^{-1}. The stretching peak between 1450 and 1400 cm^{-1} was attributed to carboxylate COO^{-}. The characteristic peak at the

1080–1030 cm^{-1} position corresponded to the C–O stretching of polysaccharides or polysaccharide-like substances. The absorption peak appearing at 870–610 cm^{-1} was caused by C–N, N–H, and C–H bending vibrations on the benzene ring. As the fermentation process progressed, the functional groups of the organic carbon components underwent succession. In the infrared spectrum, the position of the absorption peak corresponds to the type of functional group, and the intensity or area of the absorption peak is proportional to the content of the functional group. This can be used to analyze the relative changes in the content of functional groups. Therefore, peak fitting was performed to calculate the relative content of each functional group and its evolution.

In the CK treatment group, as the fermentation time advanced, the absorption peaks at 3300–3500, 2930–2850, 1650–1630, and 1080–1030 cm^{-1} all showed a fluctuating decrease, while the characteristic peaks at 1450–1400 and 870–610 cm^{-1} showed a fluctuating increase (Fig. 2.13b). The overall content of phenolic/alcoholic O–H, aliphatic C–H, benzene rings, olefinic C=C, and polysaccharide or polysaccharide-like C–O showed a downward trend, while the amount of carboxylic acid COO^- and C–H on the benzene ring showed an upward trend. This indicates that the structure of the organic carbon components gradually became more complex and stable as the fermentation progressed. In the CK fermentation between 15 and 24 days, there were basically no obvious absorption peaks at 2930–2850 cm^{-1}. On the 15th day of fermentation, the absorption peaks of phenolic/alcoholic O–H, benzene ring olefinic C=C, and polysaccharide-like C–O were significantly enhanced, with the number of benzene ring olefinic C=C functional groups reaching a peak of 25.46%, accompanied by a weakening of the carboxylic acid COO^- characteristic peak. This result may be due to the complete decomposition of lipid-like substances and the partial decomposition of carboxylic acid substances, producing a large number of small-molecule organic matter, accompanied by the synthesis of aromatic substances. The three-dimensional fluorescence results showed that the content of protein-like substances rapidly decreased from 21.53% on day 11 to 12.25%, while the content of humic-like substances rapidly increased from 37.29 to 54.69%. This indicates that while a large amount of small-molecule organic matter was decomposing, more stable humic-like substances were polymerizing, indicating that the oxidation-synthesis reaction was more intense at this time, which means that the microbial activity was higher. At this time, the pH of the material was 8.78, which is weakly alkaline, providing a suitable microenvironment for microbial growth and reproduction. Compared with the beginning of fermentation, the content of carboxylic acid COO^- was 42.28% after 24 days of fermentation, reaching the peak value in the fermentation process. It dropped to 21.47% on the 30th day of fermentation. At this time, the C–H on the benzene ring and aliphatic C-H increased from 11.88% and 0 to 17.59% and 26.95%, respectively. This indicates that in the later stage of fermentation, carboxylic acid substances were slightly decomposed, which may produce small molecular groups such as C-H, and undergo aliphatization or aromatization reactions, enhancing the activity of fulvic-like substances.

As shown in Fig. 2.13d, the overall trend of the characteristic peaks in the JJ treatment group was: the absorption peaks at 2930–2850 and 1450–1400 cm^{-1} showed

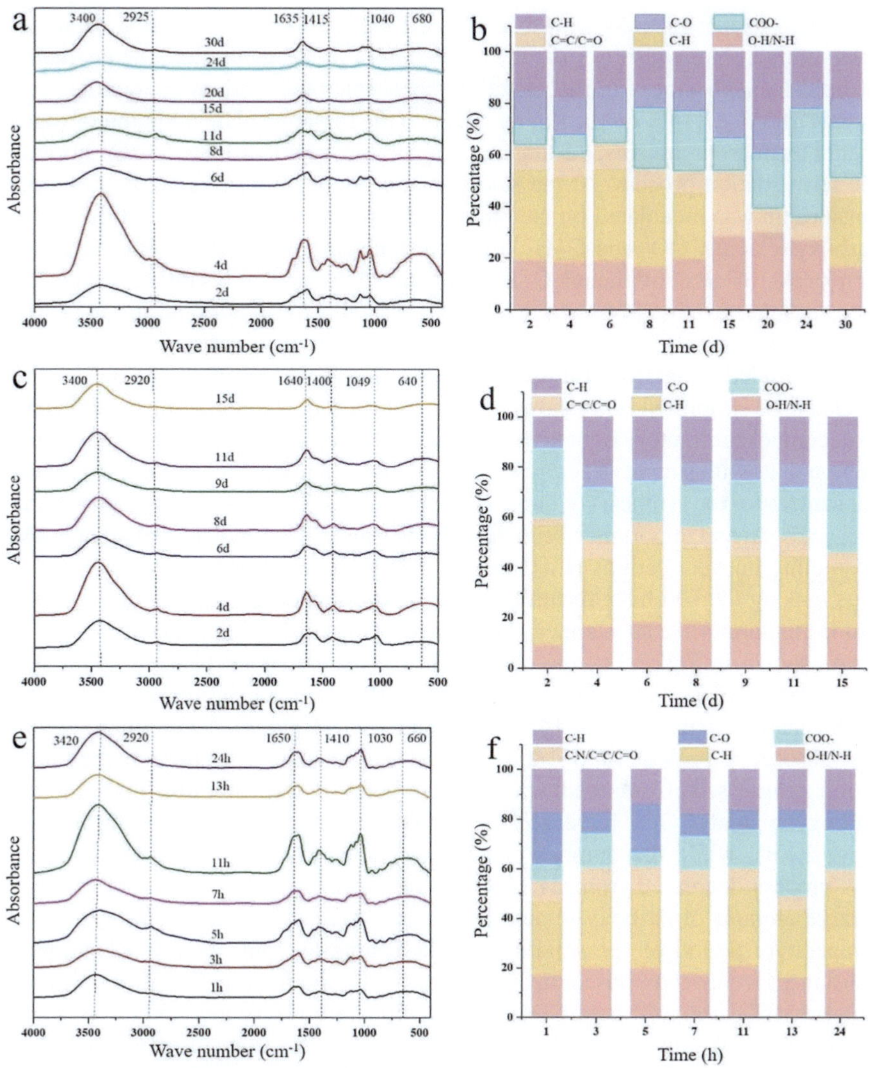

Fig. 2.13 Infrared spectra of organic carbon components and changes in the relative content of functional groups. **a** CK infrared spectrum, **b** CK functional group relative content chart, **c** JJ infrared spectrum, **d** JJ functional group relative content chart, **e** RC infrared spectrum, **f** RC functional group relative content chart

a downward trend, while the characteristic peaks at 3300–3500, 1650–1630, 1080–1030, and 870–610 cm^{-1} showed an upward trend. This indicates that the number of aliphatic methylene C–H and carboxylic acid COO^- functional groups gradually decreased during the fermentation process, while the content of alcohol/phenol O–H, aromatic olefinic C=C, polysaccharide-like C–O, and C–H on the benzene ring continued to increase. As shown in Fig. 2.13d, due to the inoculation of an acid-tolerant microbial consortium at the beginning of the fermentation, the degradation of lipid-like substances in the material was accelerated, and its content decreased from 47.85 to 27.14% on the 4th day. Due to the high-speed degradation of organic matter by microorganisms, the material temperature rapidly rose to 55 °C, shortening the fermentation process. At the same time, a lignocellulosic-degrading microbial consortium was inoculated on the 4th day of fermentation. The absorption peaks of carboxylic acid COO^- and C–H on the benzene ring weakened, while the absorption peaks of alcohol/phenol O–H, aliphatic C–H, and polysaccharide-like C–O were enhanced, but carboxylic acid COO^- was still the functional group with the highest content in JJ. This indicates that carboxylic acid substances were somewhat degraded, aromatization was weakened, and the structure on the benzene ring tended to be simplified. This shows that the lignocellulosic-degrading bacteria began to decompose refractory organic matter, producing easily degradable organic carbon components with simple structures and small molecular weights, providing precursor substances for the synthesis of humic-like substances. The carboxyl groups in humic-like acids play an important role in disease resistance. They can adsorb plant growth-promoting nutrient cations (Ca^{2+}, Mg^{2+}, K^+) with deprotonated hydroxyl groups, thus making the soil have fertility retention. The higher the content of hydrophilic groups in humic-like substances, the stronger its water-holding capacity.

As shown in Fig. 2.13f, the changes in the characteristic peaks at 3300–3500, 2930–2850, 1650–1630, and 870–610 cm^{-1} in RC were not significant. The changes in the characteristic peaks at 1450–1400 and 1080–1030 cm^{-1} were more obvious. The absorption peak at 1450–1400 cm^{-1} was significantly enhanced as the fermentation progressed. Compared with the 1st hour of fermentation, the content of carboxylic acid COO^- increased by 20.97% at the 13th hour of fermentation (Fig. 2.15f), while the number of polysaccharide-like C-O functional groups in this process decreased from 21.18 to 7.38% at the 13th hour. This indicates that a large amount of carboxylic acid substances were synthesized during the high-temperature period of fermentation, while a large amount of polysaccharide-like substances were decomposed. In the later stage of fermentation, the vibration absorption peak of carboxylic acid COO^- was significantly weakened, and the content of carboxylic acid substances fell back to 12%, while the content of alcohol/phenol O–H, aliphatic C–H, olefinic and aromatic C=C all increased. This indicates that in the later stage of fermentation, the decomposition of carboxylic acid substances to produce small-molecule organic matter was accompanied by the polymerization of aromatic substances. At the end of fermentation, the total content of alcohol/phenol O–H and aliphatic C–H functional groups was as high as 52.67%, indicating that small-molecule organic matter was the main component in the fermentation product, which was consistent with the molecular weight fractionation results.

Unlike CK, in the JJ treatment group, due to the inoculation of the acid-tolerant microbial consortium at the beginning of the fermentation, the content of aliphatic methylene C-H decreased from 47.87% on the 2nd day to 27.14% on the 4th day, while the CK group only decreased by 2.62%. At this time, the content of carboxylic acid COO^- in JJ was 20.41%, which was 2.7 times that of the CK treatment group, indicating that the inoculation of microorganisms can significantly promote the transformation of organic carbon components. At the end of fermentation, the contents of the carboxyl COO^- functional group in the three treatment groups were 21.47%, 24.72%, and 15.49%, respectively, indicating that the carboxylic acid content in the CK and JJ groups was higher than that in the RC group, with the JJ group being the highest. The sum of the content of alcohol/phenol O–H and aliphatic methylene C–H functional groups in RC was 8.82% and 11.99% higher than that of the other two groups, respectively, indicating that the content of small-molecule substances such as alcohols, phenols, and aliphatic compounds in RC was higher than that in the other two groups. The contents of functional groups such as C=C, C=O, and polysaccharide-like C–O in the three treatment groups were similar, indicating that the changing trends of aromatic compounds and polysaccharide-like substances were similar. The macromolecular substances in the CK and JJ treatment groups were higher than those in the RC group, and the small-molecule substances in the RC group were higher.

2.3.2.3 UV–Visible Spectroscopy

The UV–Visible spectra of the organic carbon components under different biological treatments (Fig. 2.14) show that as the fermentation time progresses, the absorbance of the three treatment groups decreases with increasing wavelength and eventually plateaus. All groups show a distinct absorption peak at a wavelength of approximately 280 nm. The shoulder peak here may be caused by n-π^* electronic transitions, which are related to the humic-like substances in the organic carbon components. As the content of aromatic and unsaturated conjugated double bonds increases, the UV absorption peak is enhanced. Furthermore, as fermentation time increases, the degrees of aromatization and humification continuously deepen.

$SUVA_{254}$ and $SUVA_{280}$ are spectral indicators that are positively correlated with humification and the molecular weight of organic carbon, offering greater representativeness and higher accuracy in characterizing organic carbon components. The characteristic value $SUVA_{254}$ is generally used to characterize the degree of aromatization and the level of humification of organic carbon components. It is defined as

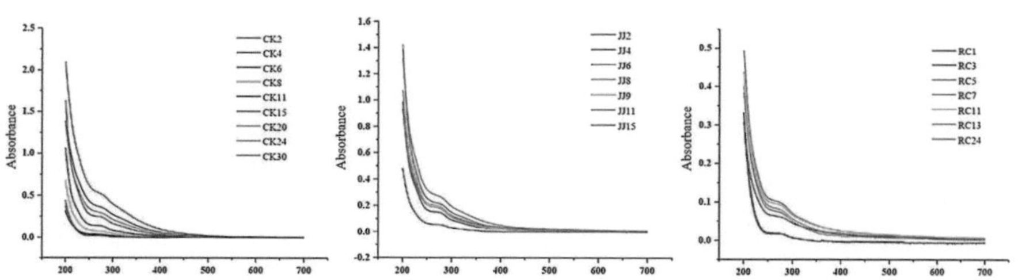

Fig. 2.14 UV–visible spectra of organic carbon components under different biological treatments

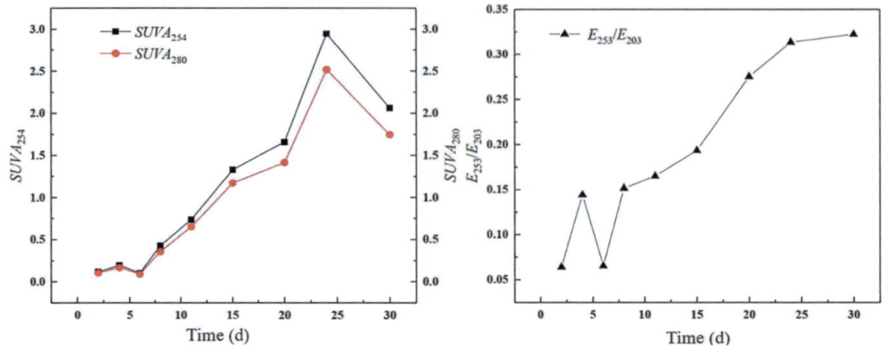

Fig. 2.15 Change patterns of $SUVA_{254}$, $SUVA_{280}$, and E_{253}/E_{203} in the CK Group

the ratio of the absorbance of organic matter at 254 nm to the DOC concentration, usually expressed as $SUVA_{254} = A_{254} * 100 /TOC$. The aromaticity and humification level of organic matter increase as this value increases. $SUVA_{280} = A_{280} * 100/TOC$, and a larger value indicates a larger molecular weight. The changes in $SUVA_{254}$ and $SUVA_{280}$ for the organic carbon components in the CK treatment group are shown in Fig. 2.15. $SUVA_{254}$ first increased and then decreased, rising rapidly from the start of fermentation to a peak value of 2.95 on day 24, indicating a gradual increase in the degree of humification. The reason for this may be that in the early and middle stages of fermentation, the metabolic activity of microorganisms was enhanced with the rise in temperature, and the hydrolysis of large-molecule organic carbon components released a large number of unsaturated aromatic compounds. These substances accumulated in the liquid phase of the material, increasing the degree of aromatization and humification of the material. As the easily utilizable organic carbon components in the material were depleted, microorganisms required energy and decomposed a small portion of the unsaturated compounds to meet their own growth and reproduction needs, causing the value of $SUVA_{254}$ to decrease. When combined with the three-dimensional fluorescence spectroscopy data, it was found that the content of protein-like substances and soluble microbial byproducts decreased as the $SUVA_{254}$ value increased, while the value increased as humic-like substances increased, with the two spectral methods corroborating each other.

The trend of $SUVA_{280}$ was consistent with that of $SUVA_{254}$, indicating that the molecular weight of the organic carbon components increased with the extension of fermentation time. The growth rate was faster in the early stage of fermentation, reaching a maximum value of 2.51 on day 24, and slightly decreasing at the end of fermentation. The reason for this may be the reduction of small-molecule organic carbon components available for microbial utilization. Overall, $SUVA_{254}$ and $SUVA_{280}$ showed an upward trend, indicating that during the fermentation process, aromaticity increased, molecular weight increased, and the degree of humification improved. Some studies have shown that the type of substituent on the benzene ring in organic carbon components is related to the $E_{253}/E_{203} = A_{253}/A_{20}$ value. This value is used to characterize the type of substituent on the benzene ring in organic matter molecules. A decrease in this value indicates that the substituents on the benzene ring are mainly aliphatic chains, while an increase indicates an increase in the number of oxygen-containing functional groups such as carbonyl, carboxyl, and hydroxyl groups in the substituents. The value of E_{253}/E_{203} showed an upward trend, indicating an increase in the number of oxygen-containing functional groups during the CK fermentation process. This value increased rapidly after the 6th day. The reason for this may be that the high temperature led to high microbial activity, which decomposed a large amount of easily degradable organic matter. The pH of the material gradually changed from weakly acidic to weakly alkaline, which was more suitable for the growth and reproduction of microorganisms, thus accelerating the microbial transformation rate. Combined with the relative changes in the content of infrared functional groups in Fig. 2.15b, it can be seen that the carboxyl content increased significantly. Therefore, the carboxyl content in the substituents on the benzene ring may be relatively high, making the types of substituents on the benzene ring more complex and the product structure more stable.

The $SUVA_{280}$ and $SUVA_{254}$ of JJ showed a continuous upward trend (Fig. 2.16). In the first 9 days of fermentation, the values of both increased rapidly, and then the growth rate slowed down. The reason may be that the inoculation enhanced the conversion rate of small-molecule organic matter to large-molecule humic substances, improving the degree of aromatization and humification of the material. The subsequent slowing of the growth rate was because in the later stage of fermentation, the temperature began to drop, and the activity of microorganisms decreased, thus slowing down the conversion rate. Parallel factor analysis data showed that during the JJ fermentation process, the content of small-molecule organic matter component C1, soluble microbial byproducts, in the material decreased from 46.01 to 39.56% at the end of fermentation, and the protein-like substances in C4 decreased by 34.60%, while the large-molecule organic matter components C2 + C3, humic-like substances, increased from 7.89 to 48.95%. This confirms that fermentation is a process in which small-molecule organic matter is decomposed, transformed, and polymerized by microorganisms to form large-molecule, refractory organic matter. The value of $SUVA_{254}$ for JJ increased to 1.53. As fermentation proceeded, the structures of aromatic substances and unsaturated conjugated double bonds increased. These two structures can further produce phenol hydroxyl, carboxyl, and carbonyl groups, thus causing the value of E_{253}/E_{203} to show an upward trend, rising from an initial 0.15–0.26 at the end of fermentation. This value fluctuated greatly during the fermentation process, indicating that the aerobic reaction of organic carbon components was intense in the early stage of fermentation, the number of oxygen-containing substituents on the aromatic ring increased rapidly, and the content of aliphatic chains in the benzene ring substituents decreased, possibly being converted into more stable oxygen-containing substituents such as carbonyl and carboxyl groups, or being aerobically decomposed and transformed by microorganisms. Infrared spectroscopy studies showed that at the end of fermentation, the content of aliphatic methylene C-H decreased significantly from 47.85 to 24.76%, while the content of alcohol hydroxyl O–H and carbonyl C=O increased, indicating that the substituent structure on the

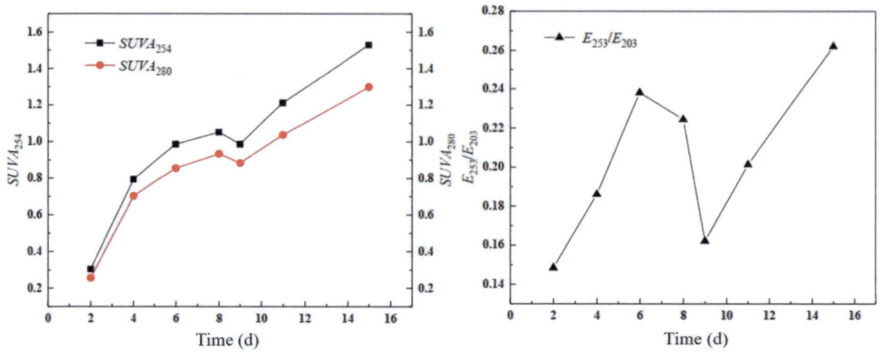

Fig. 2.16 Change patterns of $SUVA_{254}$, $SUVA_{280}$, and E_{253}/E_{203} in the JJ Group

aromatic ring tended to become more complex. Combining these three characteristic values, it was found that as the fermentation process proceeded, the degree of humification increased, the molecular weight increased, and the benzene ring structure tended to become more complex and stable.

The three characteristic values of RC, $SUVA_{254}$, $SUVA_{280}$, and E_{253}/E_{203}, changed consistently, decreasing rapidly in the first 3 h, rising sharply after 5 h, and then starting to gradually decrease to a stable level after the 7th hour. All three reached their maximum values at 7 h. The changes in $SUVA_{254}$ and $SUVA_{280}$ were not large, increasing by 0.07 and 0.06, respectively, while the value of E_{253}/E_{203} decreased slightly (Fig. 2.17). This indicates that the degree of humification of the material was not high, and it mainly contained small-molecule organic carbon components, with a higher number of aliphatic chains in the benzene ring substituents. This may be due to the inoculation at the beginning of fermentation, where microorganisms decomposed and produced a large amount of small-molecule organic carbon components. Combined with three-dimensional fluorescence spectroscopy, it is known that the content of both tyrosine and tryptophan components increased between 1 and 5 h. As the temperature rose, microbial activity was enhanced. Due to the large accumulation of small-molecule organic carbon components in the first 5 h, the small-molecule organic carbon components rapidly polymerized after 5 h to form humic-like substances with larger molecular weights, causing the three characteristic values to increase rapidly. The highest temperature of 71 °C was reached within 7 h, and some microorganisms in the material became inactive or entered a dormant state, leading to a decrease in their ability to oxidize and decompose organic matter, and the three characteristic values decreased slightly. The data from three-dimensional fluorescence showed that at the end of fermentation, the total content of soluble microbial byproducts and protein-like substances was as high as 81.65%, and the pH was 5.29, in a weakly acidic state, indicating that the content of organic acids in the material was high, and a large amount of organic acids had not yet been transformed and polymerized to form humic-like substances. In addition, the infrared functional group data showed that the content of aliphatic methylene C-H was 32.37%, while

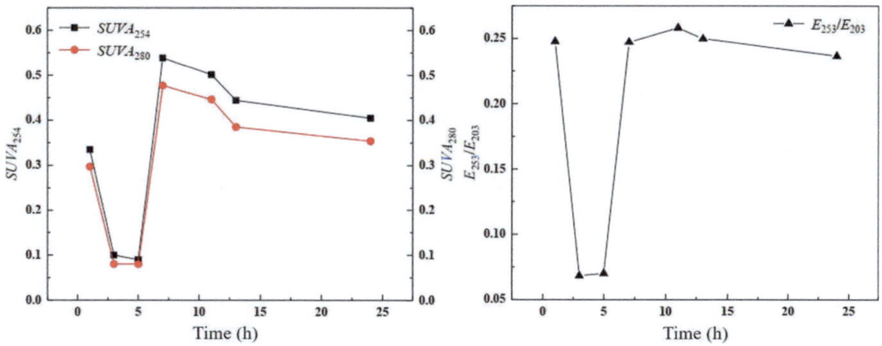

Fig. 2.17 Change patterns of $SUVA_{254}$, $SUVA_{280}$, and E_{253}/E_{203} in the RC Group

that of aromatic C=O was only 7.40%, and carboxyl COO^- was only 15.49%. This indicates that the RC fermentation product is mainly composed of small-molecule organic carbon components.

In the CK group, the three characteristic values $SUVA_{254}$, $SUVA_{280}$, and E_{253}/E_{203} rose from 0.12, 0.11, and 0.06 to 2.06, 1.74, and 0.32, respectively. In the JJ group, $SUVA_{254}$, $SUVA_{280}$, and E_{253}/E_{203} rose from 0.30, 0.26, and 0.15 to 1.53, 1.29, and 0.26, respectively. In the RC group, $SUVA_{254}$, $SUVA_{280}$, and E_{253}/E_{203} rose from 0.34, 0.30, and 0.25 to 0.40, 0.35, and 0.24, respectively. The order of magnitude of these three characteristic parameters for the three treatment groups was CK > JJ > RC. This indicates that CK had the highest degree of humification, the largest molecular weight, and a more complex aromatic ring structure, followed by the JJ treatment group, and finally RC, which is consistent with the results of molecular weight fractionation. The reasons for this difference are, first, the different fermentation times, which lead to different degrees of maturation and aromatization; and second, the inoculation of microbial consortia, which leads to differences in the changes of these three characteristic values.

2.3.2.4 Response Relationship Between Organic Carbon Forms and Environmental Factors

To investigate the key environmental factors affecting the transformation of organic carbon components under different biological treatment methods, a correlation analysis was performed (Fig. 2.18). In the CK treatment group, a highly significant negative correlation ($p < 0.01$) was found between OM (organic matter) and V (humic-like acid) and $SUVA_{280}$. A highly significant positive correlation ($p < 0.01$) was observed between OM and I (tyrosine-like), II (tryptophan-like), and IV (soluble microbial byproducts). A significant positive correlation ($p < 0.05$) was found with Fatty C–H (aliphatic methylene). pH showed a highly significant negative correlation with I, II, IV, and aliphatic methylene (Fatty C–H), and a highly significant positive correlation with V and $SUVA_{280}$. TOC had a significant negative correlation with carboxyl

COO^-, a highly significant positive correlation with I, II, and IV, a highly significant negative correlation with V and $SUVA_{280}$, and a significant positive correlation with aliphatic methylene. Therefore, in the CK group, OM, pH, and TOC were the key environmental factors influencing the transformation of organic carbon components during food waste fermentation. A highly significant correlation exists between OM, pH, and $SUVA_{280}$, indicating that regulating the organic matter content or pH in the material can affect the molecular weight of the organic carbon components. For instance, increasing the pH may have the potential to promote the formation of macromolecular humic carbon components, while increasing the organic matter content in the material may have the potential to promote the formation of protein-like components. Component I showed a highly significant positive correlation with II, IV, and aliphatic methylene C–H, indicating that the formation of component I has a certain similarity to the formation of components II, IV, and aliphatic methylene. Component V and $SUVA_{280}$ both showed a highly significant negative correlation with components I, II, and IV. Furthermore, V had a significant positive correlation with carboxyl COO^- and a highly significant positive correlation with $SUVA_{280}$. Research indicates that the formation of humus mainly involves two stages. The first stage is the decomposition of complex, large-molecule organic matter into simpler, small-molecule organic matter (such as proteins, amino acids, and polypeptides) by microorganisms. In the second stage, microorganisms polymerize a portion of the small-molecule organic carbon components into humic-like substances, while another portion is oxidized and decomposed to produce substances like CO_2, H_2O, NH_3, and inorganic salts. The study by Zhu et al. also showed that fermentation is a process of transforming low-molecular-weight organic carbon into high-molecular-weight organic carbon components (Zhu et al. 2020). Therefore, the reduction of small-molecule organic carbon components like tyrosine and tryptophan is highly likely to be due to their transformation into large-molecule humic-like substances.

As shown in Fig. 2.19, in the JJ treatment group, T (temperature) showed a significant positive correlation with C1 (soluble microbial byproducts and tryptophan). OM showed a significant positive correlation with C3 (protein-like substances), a highly significant negative correlation with C2 (fulvic-like acid substances) and C4 (humic-like acid substances), and a significant negative correlation with $SUVA_{280}$. TOC had a highly significant positive correlation with C3, a highly significant negative correlation with C2 and C4, and a significant negative correlation with $SUVA_{280}$. pH showed a highly significant positive correlation with C2 and C4, a highly significant negative correlation with C3, and a significant positive correlation with $SUVA_{280}$. Therefore, in the JJ treatment, T, OM, TOC, and pH were the controlling factors affecting the transformation of organic carbon components of different molecular weights. Increasing the values of T, OM, and TOC may have a positive contribution to promoting the production of small-molecule organic carbon components. A high pH has a significant effect on promoting the formation of large-molecule humic-like substances, which is consistent with a previous report. C3 showed a significant positive correlation with aliphatic methylene C–H, a significant negative correlation with C4, polysaccharide-like C–O, and C–H on the aromatic ring, and a highly significant negative correlation with C2. C2 had a significant negative correlation with C1, a

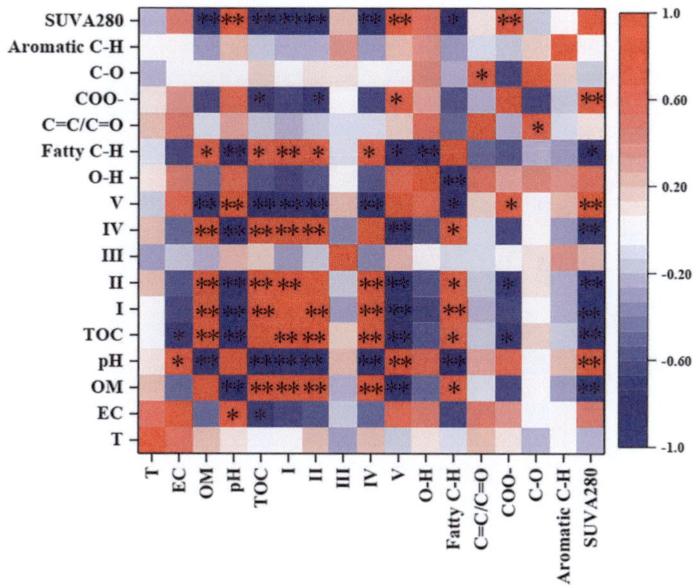

Fig. 2.18 Correlation analysis of organic carbon components and physicochemical factors in the CK Group T (temperature), EC (electrical conductivity), OM (organic matter), TOC (total organic carbon), I (tyrosine-like), II (tryptophan-like), III (fulvic-like acid substance), IV (soluble microbial byproduct), V (humic-like acid substance), O–H (phenolic/alcoholic hydroxyl), Fatty C-H (aliphatic methylene), C=C/C=O (olefinic, aromatic C=C), COO− (Carboxyl COO−), C–O (Polysaccharide-like C–O), Aromatic C–H (C–H on benzene ring), $SUVA_{280} = A_{280}*100/TOC$. *Note* * indicates $p < 0.05$, ** indicates $p < 0.01$

highly significant negative correlation with C3, a highly significant positive correlation with C4, and a significant positive correlation with $SUVA_{280}$. This indicates that the formation patterns of C2 and C4 organic carbon components have a certain similarity, and the large-molecule C2 and C4 organic carbon components are very likely derived from the decomposition and transformation of C1 and C3 organic carbon components.

As shown by the correlation heatmap analysis in Fig. 2.20, a significant negative correlation exists between T and V. pH showed a significant negative correlation with III and a highly significant positive correlation with $SUVA_{280}$. Component III had a significant negative correlation with $SUVA_{280}$. This indicates that T and pH are important factors affecting the transformation of organic carbon components in the RC treatment group. Components III and V are collectively referred to as humic-like substances; therefore, T and pH have a significant impact on the formation of large-molecule humic-like substances. Component II showed a highly significant negative correlation with component V, indicating a certain relationship between the decomposition of tryptophan and the synthesis of humic-like acid.

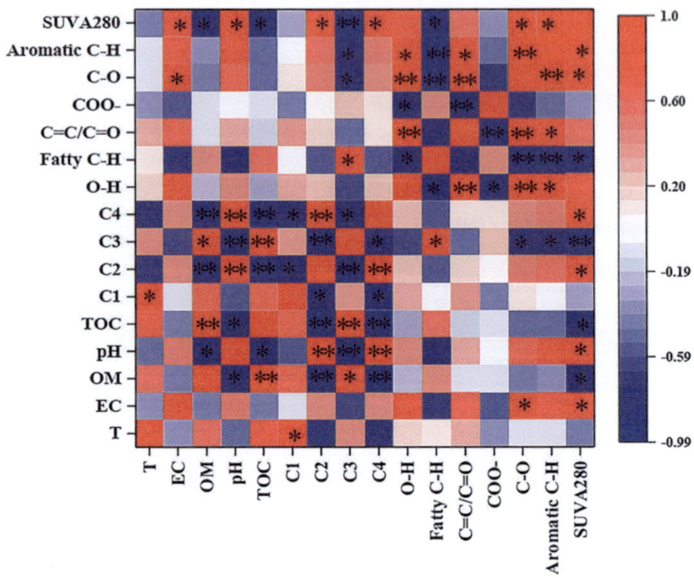

Fig. 2.19 Correlation analysis of organic carbon components and physicochemical factors in the JJ Group T (temperature), EC (electrical conductivity), OM (organic matter), TOC (total organic carbon), I (tyrosine-like), II (tryptophan-like), III (fulvic-like acid substance), IV (soluble microbial byproduct), V (humic-like acid substance), O–H (Phenolic/alcoholic hydroxyl), Fatty C–H (Aliphatic methylene), C=C/C=O (Olefinic, aromatic C=C), COO– (Carboxyl COO–), C–O (Polysaccharide-like C–O), Aromatic C–H (C–H on benzene ring), $SUVA_{280} = A_{280}*100/TOC$. *Note* * indicates $p < 0.05$, ** indicates $p < 0.01$

From the above research, the following conclusions can be drawn: (1) As fermentation proceeded, the organic matter in all experimental groups showed a downward trend. (2) The spectral characterization results indicated that the CK group had the highest aromaticity, the largest molecular weight, and the most humic-like substances. The JJ group contained a higher amount of COO^-, and its organic carbon components were rich in carboxylic acid substances and fulvic-like acid substances. The RC group had the lowest aromaticity and the smallest molecular weight compared to the CK and JJ groups. (3) The correlation analysis results showed that pH had a significant or highly significant positive correlation with $SUVA_{280}$, indicating that pH has a significant impact on the molecular weight of organic carbon components, and a high pH may promote the formation of large-molecular-weight organic carbon components. In the CK and JJ treatment groups, organic matter and total organic carbon showed a significant or highly significant positive correlation with protein-like substances and a highly significant negative correlation with humic-like substances, indicating that organic matter and total organic carbon are major factors affecting the formation of organic carbon components, and increasing the organic matter content in the material may have the potential to promote the

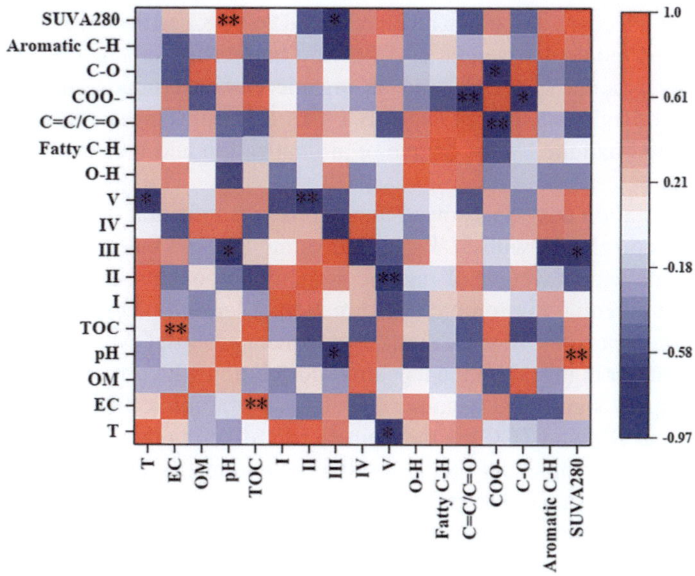

Fig. 2.20 Correlation analysis of organic carbon components and physicochemical factors in the RC Group T (temperature), EC (electrical conductivity), OM (organic matter), TOC (total organic carbon), I (tyrosine-like), II (tryptophan-like), III (fulvic-like acid substance), IV (soluble microbial byproduct), V (humic-like acid substance), O–H (phenolic/alcoholic hydroxyl), Fatty C–H (aliphatic methylene), C=C/C=O (olefinic, aromatic C=C), COO− (Carboxyl COO−), C–O (polysaccharide-like C–O), Aromatic C–H (C–H on benzene ring), $SUVA_{280} = A_{280}*100/TOC$. *Note* * indicates $p < 0.05$, ** indicates $p < 0.01$

formation of protein-like substances. In the RC group, the formation of humic-like substances was mainly affected by temperature and pH.

To explore the response relationship between the transformation of organic carbon components and physicochemical factors under different biological treatment methods and to reveal the key factors affecting the transformation of organic carbon groups, the physicochemical factors were measured, the organic carbon components were characterized by spectroscopy, and a correlation analysis was performed, leading to the following conclusions:

(1) The fermentation temperature of the CK and JJ treatment groups was maintained at 55 °C for at least 3 days, eliminating parasites and pathogens in the material and meeting the sanitization standard. The pH change showed that inoculating with an acid-tolerant microbial consortium could alleviate the acidification of the material. The organic matter in all three fermentation groups showed a downward trend. TOC continuously decreased in CK and JJ as fermentation progressed, while it slightly increased in RC, which may be due to the accumulation of water-soluble organic matter in the material. The electrical conductivity and germination rate values indicated that the fermentation products had no toxic effects on plants.

(2) The spectral characterization results showed that CK had the highest aromaticity, the largest molecular weight, and the most humic-like substances. JJ contained a higher amount of COO^-, and its organic carbon components were rich in carboxylic acid substances and fulvic-like acid substances. Compared to CK and JJ, RC had the lowest degree of aromatization and the smallest molecular weight.
(3) The correlation analysis showed that pH had a significant or highly significant positive correlation with $SUVA_{280}$, indicating that pH has a significant impact on the molecular weight of organic carbon components, and a high pH may have the potential to promote the formation of large-molecular-weight organic carbon components. In the CK and JJ treatment groups, OM and TOC showed a significant or highly significant positive correlation with protein-like substances and a highly significant negative correlation with humic-like substances, indicating that OM and TOC are the main factors affecting the formation of organic carbon components, and increasing the organic matter content in the material may have the potential to promote the formation of protein-like substances. In RC, the formation of humic-like substances was mainly affected by temperature and pH.

2.4 Microenvironmental Changes During Aerobic Fermentation of Food Waste

High-throughput sequencing technology can be used to elucidate the composition and succession patterns of the microbial community during the fermentation process. By combining the physicochemical properties of the fermentation products, changes in organic carbon morphology under different biological treatment conditions, and the influence of key environmental factors on functional microorganisms and the transformation of different organic carbon components, the response relationship between functional microorganisms and organic carbon components can be revealed.

2.4.1 Bacterial Community Succession During Fermentation

2.4.1.1 Analysis of Bacterial Community Diversity

Microbial diversity analysis can reveal the types and numbers of microorganisms involved in the fermentation process, which helps us understand the roles and mechanisms of these microorganisms during fermentation. Alpha diversity analysis of the bacterial community is an important evaluation index in the aerobic composting process, as it helps to understand the distribution of bacterial species and numbers.

High-throughput sequencing of microorganisms was performed on samples from the warming, thermophilic, and cooling phases of the three fermentation processes: CK, JJ, and RC (Table 2.3). The samples for CK were CK1, CK8, CK11, CK17, and CK30. JJ included JJ1, JJ4, JJ6, JJ9, and JJ15. RC had RC1, RC3, RC7, RC11, and RC24 (the numbers represent the fermentation time: d for CK and JJ, h for RC).

Indices such as Shannon, Simpson, Chao1, ACE, and Coverage were selected for Alpha diversity analysis. Among them, Shannon and Simpson are indices for assessing species diversity. A higher Shannon value indicates higher microbial diversity and a more even community distribution, while a higher Simpson value indicates lower microbial community diversity. ACE and Chao1 represent microbial species abundance; the values of these two indices are directly proportional to the richness of the microbial community. The optimized sequence numbers for the CK, JJ, and RC groups were 253,210, 213,148, and 265,496, respectively. Venn diagram analysis (Fig. 2.21) shows that the three treatment groups shared 1212 bacterial OTUs. The number of OTUs for CK, JJ, and RC were 610, 436, and 902, respectively, with unique bacterial OTU proportions of 13.70%, 3.55%, and 41.01%. The number of OTUs common to all three was 230. It is evident that RC had more OTU types and greater differences. CK and JJ shared 101 bacterial OTUs, CK and RC shared 113 bacterial OTUs, and JJ and RC shared 62 bacterial OTUs, indicating that the bacterial community structures of JJ and CK were relatively similar.

Table 2.3 shows that the Coverage of effective bacterial sequences for the three groups of fermentation samples was all greater than 99%, which can represent the true situation of the bacterial community in the samples. As the fermentation process

Table 2.3 Bacterial diversity index table

Sample No.	Shannon	Simpson	ACE	Chaol	Coverage
CK1	2.35	0.14	94.69	98.43	0.999
CK8	1.82	0.37	329.22	214.71	0.997
CK11	1.85	0.34	91.58	87.11	0.999
CK17	3.78	0.07	348.57	375.71	0.997
CK30	4.70	0.02	500.51	516.56	0.996
JJ1	3.72	0.05	233.80	225.60	0.998
JJ4	2.20	0.24	138.70	140.25	0.999
JJ6	1.19	0.64	188.36	184.30	0.998
JJ9	3.32	0.08	211.28	234.00	0.998
JJ15	4.15	0.04	400.68	401.03	0.996
RC1	2.72	0.23	567.77	550.01	0.992
RC3	2.07	0.47	721.31	706.00	0.990
RC7	1.48	0.62	451.50	472.29	0.995
RC11	1.74	0.53	464.95	443.59	0.994
RC24	2.06	0.20	223.20	142.13	0.998

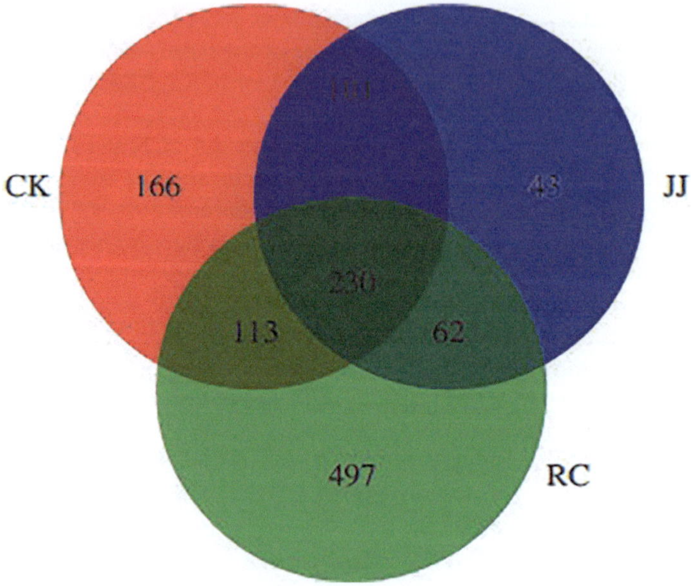

Fig. 2.21 Venn diagram at the bacterial OTU level

progressed, the Shannon index value of the CK treatment group decreased from an initial 2.35 to 1.85 during the thermophilic phase, and then rose to 4.70 by the end of fermentation. The Simpson value increased from 0.14 to 0.34 during the thermophilic phase and then decreased to 0.02 by the end of fermentation, indicating that the overall bacterial community diversity in CK showed an upward trend. The higher temperature during the thermophilic phase caused some microorganisms to enter a dormant or dead state, leading to a decrease in diversity. The bacterial community diversity was greatest at the end of fermentation. The reason for this may be that during the thermophilic phase of fermentation, microorganisms decomposed a large amount of organic matter, causing a large accumulation of nutrients in the material that are easily absorbed and utilized by microorganisms. In the later stage of fermentation, the temperature decreased and the pH increased, making the microenvironment in the material more suitable for microbial growth and reproduction. At the beginning of fermentation, the Shannon index for JJ was 3.72 while for CK it was 2.35, indicating that inoculating with a microbial consortium can significantly increase the microbial community diversity in the material, with a more significant effect in the early stage of fermentation. The trend of bacterial diversity index changes in the RC treatment group was different from the other two groups. As fermentation progressed, the Shannon index decreased, and the Simpson index increased, indicating a decrease in bacterial community diversity. The reason may be that the temperature in this treatment group was too high during the thermophilic phase (about 71 °C), leading to the inactivation or death of some bacteria. In addition, due to the accumulation of organic acids in the material, the pH remained around 5, in a weakly acidic state, which inhibited the growth of the colonies and reduced diversity.

Both the ACE index and the Chao1 index can represent colony richness; the higher the value, the higher the colony abundance. In the CK and JJ treatment groups, the trends of the two index values were consistent, both showing a fluctuating upward trend, indicating that the richness of the microbial community fluctuated up and down, ultimately showing an increasing trend. In CK, the two index values dropped to their lowest on day 11, possibly because the fermentation was in the thermophilic phase at this time, the material temperature was high, affecting microbial reproduction, and the microbial richness was low. By the end of fermentation, the temperature decreased, dominant bacteria underwent succession, the ACE and Chao1 indices gradually increased, and the microbial community abundance gradually rose. In JJ, the ACE and Chao1 indices decreased to a certain extent on the 4th day of fermentation. Compared with CK, the decrease was smaller, possibly because the JJ treatment group was inoculated with an acid-tolerant microbial consortium in the early stage of fermentation, which increased the microbial community abundance in the material. Then, on the 6th day, the ACE and Chao1 index values began to rise, because a lignocellulosic-degrading microbial consortium was inoculated on the 4th day of fermentation. The inoculated exogenous microorganisms adapted to the microenvironment in the material, which increased the microbial colony abundance, indicating that the exogenous inoculation of the consortium increased the abundance of the microbial community. In the RC treatment group, the ACE and Chao1 index values showed a fluctuating downward trend, indicating a decrease in microbial abundance.

2.4.1.2 Composition and Dynamic Succession of Bacterial Communities

Changes at the Bacterial Community Phylum Level: Microorganisms, as the core element of fermentation, are a key factor influencing the transformation of organic carbon components. The metabolic activity of bacteria is the main reason for the temperature increase and organic matter degradation in aerobic fermentation. Therefore, it is crucial to reveal the succession of the bacterial community structure in the material during the fermentation process. High-throughput sequencing technology can reveal the structural characteristics of the microbial community during the fermentation process and its impact on the transformation of organic carbon components with high resolution and precision. The changes in the bacterial community at the phylum level under different biological treatment methods are shown in Fig. 2.22. There were 10 phyla with a relative abundance of > 1% at the bacterial community phylum level in the samples. In all fermentation processes, the main dominant phyla were *Firmicutes*, *Proteobacteria*, *Bacteroidota*, and *Actinobacteriota*. At the beginning of fermentation in the CK group, *Firmicutes* accounted for over 90% of the total phyla, making it the absolute dominant phylum. In the middle stage of fermentation, the abundance of *Proteobacteria* increased, and the main dominant phyla evolved to be *Proteobacteria* and *Firmicutes*. In the later stage of fermentation, the abundance of *Firmicutes* significantly decreased, while the abundances of *Proteobacteria*, *Bacteroidota*, and *Actinobacteriota* began to gradually increase. *Proteobacteria* increased to 36.61%, and the other two phyla were 4.93% and 2.96%, respectively. At this time, *Myxococcota* appeared in the material with an abundance of 1.58%. The dominant phyla gradually evolved to *Firmicutes*, *Proteobacteria*, *Actinobacteriota*, and *Bacteroidota*. By the end of fermentation, *Firmicutes* had decreased to 32.23%, a 66.95% decrease in abundance compared to the initial stage of fermentation. However, *Firmicutes* still maintained its position as a dominant phylum. At the same time, other dominant phyla such as *Deinococcota*, *Chloroflexi*, *Desulfobacterota*, *Patescibacteria*, and *Gemmatimonadota* increased, with abundances of 4.04%, 1.97%, 1.40%, 0.72%, and 1.51%, respectively. This indicates that the dominant phyla in the CK fermentation process underwent succession and their diversity increased.

In the JJ fermentation process, at the initial stage, the abundances of the *Firmicutes* and *Proteobacteria* phyla were 45.79% and 47.37%, respectively, making them the main dominant phyla. By day 4 of fermentation, the abundance of the *Firmicutes*

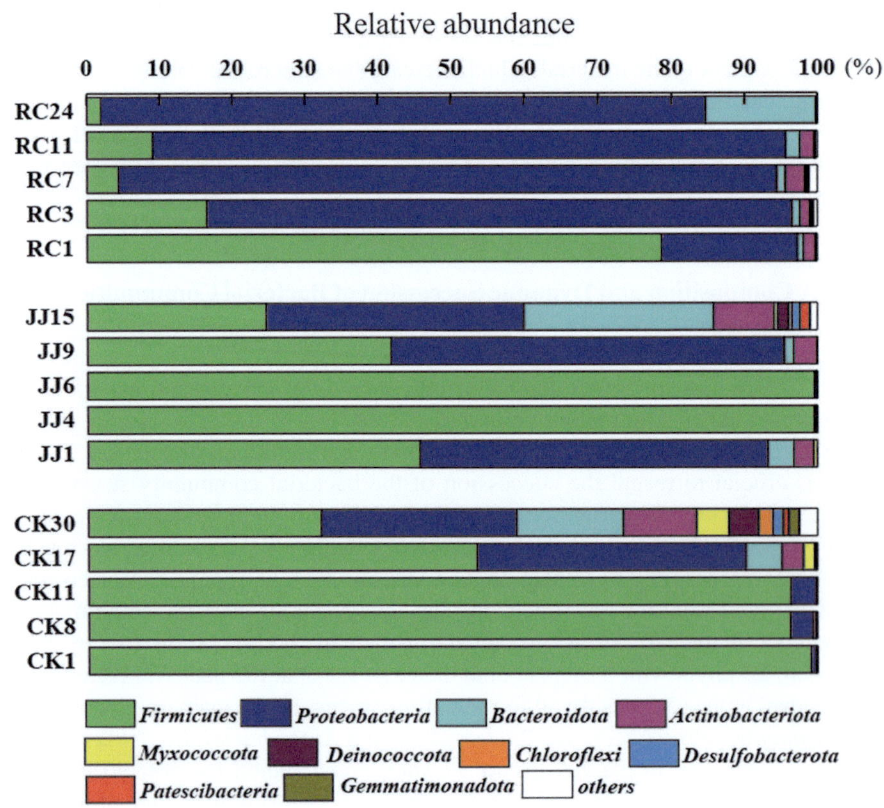

Fig. 2.22 Relative abundance at the bacterial phylum level

phylum increased to 99.57%, becoming the absolute dominant phylum. After experiencing the thermophilic phase, its abundance rapidly decreased. By the end of fermentation, the dominant phyla had evolved to *Firmicutes*, *Proteobacteria*, *Bacteroidota*, and *Actinobacteriota*, with abundances of 24.76%, 35.15%, 25.86%, and 8.27%, respectively. Unlike the CK treatment group, at the end of the JJ fermentation, the abundances of phyla such as *Myxococcota*, *Deinococcota*, *Chloroflexi*, *Desulfobacterota*, *Patescibacteria*, and *Gemmatimonadota* were lower. This may be because the inoculation of the microbial consortium increased the abundance of the dominant phyla.

In RC, the dominant phyla evolved from *Firmicutes*, *Proteobacteria*, *Bacteroidota*, and *Actinobacteriota* at the beginning of fermentation, with abundances of 78.63%, 18.62%, 0.85%, and 1.64%, respectively, to *Firmicutes*, *Proteobacteria*, and *Bacteroidota* at the end of fermentation, with abundances of 1.94%, 82.77%, and 15.10%, respectively. The abundance of the dominant phyla changed significantly.

The main reason for the differences in bacterial community succession under different aerobic fermentation methods is, on the one hand, that the inoculation of exogenous microbial consortia in JJ and RC affected the species and abundance of the indigenous microbial community in the material. Inoculation stimulates the entire fermentation microbial community, acting as a microbial catalyst. On the other hand, microorganisms use the organic carbon components in the material as carbon and nutrient sources. The organic carbon components in immature fermentation material are mainly composed of total sugars (including hemicellulose or cellulose), phenolic substances, amino acids, peptides, and other easily biodegradable compounds. The main components of organic carbon differ at different fermentation stages, leading to differences in the reproduction rates of bacteria that use different organic matter as nutrient sources. Firmicutes is a type of fermenting bacteria that can utilize a wide range of substrates and can directly metabolize large-molecule organic matter (such as proteins, starch, and cellulose) as substrates. Its abundance is high during the warming phase, making it a dominant phylum in the material. As the protein-like substances in the material were depleted, the abundance of *Firmicutes* decreased, and the abundances of *Actinobacteriota*, *Bacteroidota*, and *Proteobacteria* began to increase and evolve into the main dominant phyla. However, when the nutrient content in the material is high, the reproduction rate of *Actinobacteriota* is relatively slow. Furthermore, *Actinobacteriota* is the main phylum for degrading cellulose and plays a key role in the degradation of rice husks. The abundance of the *Actinobacteriota* phylum in RC was the lowest, which led to a smaller change in its OM content. Finally, the microenvironments in the material, such as temperature and pH, differed under the different aerobic fermentation methods. This is consistent with the research by Awasthi et al., which states that during the fermentation process, differences in environmental conditions and nutrients at each stage cause changes in bacterial diversity and abundance, leading to the successive appearance of specific microbial communities (Awasthi et al. 2018).

Changes at the Bacterial Community Genus Level: A heatmap was used to further analyze the microbial community structure. Figure 2.23 shows the expression of different species at the genus level across various samples. The clustering relationships on the heatmap can reflect the relationships between species. To reduce the impact of noise data, a threshold of Others < 3% was set, and a cluster analysis was performed on the top 25 most abundant genera. Each column represents a sample, and each row represents a classified species. Bacterial genera with higher abundance in the corresponding sample are distinguished by red, while those with lower abundance are distinguished by green. The microbial clusters were mainly divided into two major categories. The first major category was further divided into 3 sub-categories. The first sub-category mainly included 15 genera: *Fermentimonas*, *unclassified_f__Rhizobiaceae*, *Flavobacterium*, *Moheibacter*, *Enterococcus*, *Pseudoxanthomonas*, *Escherichia-Shigella*, *Paenibacillus*, *Aneurinibacillus*, *Brevibacillus*, *Sporosarcina*, *Lysinibacillus*, *Chelativorans*, *Ureibacillus*, and *Bacillus*. The second sub-category included the following 4 genera: *unclassified_f__Enterobacteriaceae*, *Acinetobacter*, *Chryseobacterium*, and *Pseudomonas*. The third sub-category included *unclassified_c__Bacilli* and *Pediococcus*. The second major category mainly consisted of *Enterobacter*, *Weissella*, *Leuconostoc*, and *Lactobacillus*.

In the CK group, the dominant genera at the initial stage of fermentation were mainly *Lactobacillus* (70.30%) and *Weissella* (18.26%). Studies have found that most lactic acid bacteria in the *Lactobacillus* genus survive under conditions with a pH of less than 5. They can decompose lignocellulose to produce lactic acid, and their life activities are less affected by acid accumulation. Due to the high abundance and vigorous life activities of the *Lactobacillus* genus in the early stage of fermentation, a large amount of lactic acid was produced and accumulated in the material,

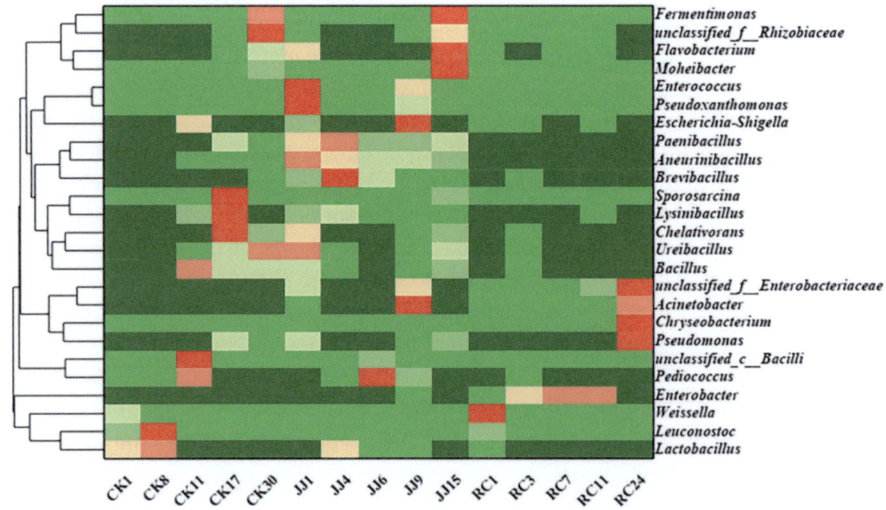

Fig. 2.23 Clustering distribution of dominant species in each treatment group

leading to a decrease in pH and putting the material in a weakly acidic state. Furthermore, *Lactobacillus* can increase the content of available inorganic salt ions such as nitrogen, phosphorus, and potassium in the fermentation product, leading to an increase in the electrical conductivity of the fermentation samples. As the fermentation progressed, small-molecule organic acids were gradually mineralized to produce CO_2 or polymerized to produce large-molecule humic-like substances. The content of small-molecule organic acids in the material decreased, and the abundance of microbial genera that use them as a nutrient source subsequently declined. The main dominant genus at the end of fermentation was succeeded by *Bacillus*, which can degrade cellulose to form humic-like substances.

The main dominant genera in the warming phase of JJ were *Lactobacillus* (66.88%), *Brevibacillus* (8.90%), *Bacillus* (5.85%), and *Pediococcus* (5.30%). During the thermophilic phase, the abundance of *Pediococcus* (a homofermentative lactic acid bacterium genus) rose to 16.35%, continuing its status as a dominant species. *Pediococcus* can accelerate the fermentation process by preventing acetic acid production and stimulating indigenous microorganisms in the material. The abundance of *Acinetobacter* was 28.68%, becoming one of the main dominant genera during this period. The abundance of *Enterobacter* rose to 5.85%, making it another dominant genus of this period. During the cooling phase, the dominant genera increased to include *Pseudomonas*, *Moheibacter*, and *Bacillus*, with abundances of 17.94%, 11.79%, and 7.35%, respectively. *Pseudomonas* has a good ability to degrade lignin and cellulose, and *Bacillus* has a strong ability to hydrolyze pectin and cellulose. Both genera accelerated the degradation of cellulose, reduced the toxicity of the fermented organic fertilizer, and improved the stability of the fermentation. Furthermore, *Bacillus* promoted the enrichment of inorganic phosphorus in the material, increasing the soluble inorganic salt ions in the material, which led to an increase in electrical conductivity. In the warming phase of JJ, the abundance of *Lactobacillus* (66.88%) was lower than that of *Lactobacillus* (70.30%) in CK during the same period, and the pH value of JJ was higher than that of CK. This may be because the acid-tolerant microbial consortium inoculated at the beginning of the JJ fermentation inhibited the growth and reproduction of Lactobacillus, alleviating the acidification in the material.

The dominant genera in RC at the initial stage of fermentation were *Lactobacillus* (18.61%), *Enterobacter* (10.39%), and *Weissella* (46.19%), with *Weissella* being absolutely dominant. Weissella is a heterofermentative lactic acid bacterium that produces acetic acid when degrading organic matter. The accumulation of a large amount of acetic acid in the material caused the pH to decrease. In the middle stage of fermentation, the abundance of *Enterobacter* remained above 68%. By the end of fermentation, the dominant genera in the material had evolved to *Pseudomonas*, *Acinetobacter*, and *Chryseobacterium*, with an abundance ratio of 3.3:1.5:1.0 for the three genera. The results at the bacterial genus level for RC indicate that *Enterobacter* played a key role throughout the entire fermentation cycle and can increase the fermentation temperature of the material, accelerate the decomposition rate of organic matter, and lower the pH. Therefore, the temperature during the high-temperature phase of RC was higher than that of the other two groups, and the microenvironment of the material was in a weakly acidic state. Both *Pseudomonas* and *Chryseobacterium* have the ability to degrade cellulose, degrading the lignocellulose in the material and providing a good foundation for the polymerization of humic substances.

Principal Coordinate Difference Analysis of Bacterial Communities: Principal Coordinate Analysis (PCoA) was used to compare the differences in bacterial communities under different biological treatment methods. Sample points that are close to each other in the PCoA plot indicate that the sample compositions are relatively similar. PCoA analysis was performed on the three treatment groups. As shown in Fig. 2.24, the contribution rates of PCA1 and PCA2 were 31.60% and 16.04%, respectively, and the sum of the main contribution rates was 47.64%. CK1, CK8, and JJ1 were far apart. CK11 and JJ4 were relatively close, and CK17, CK30, and JJ15 were relatively close. This indicates that there was a large difference in the bacterial community composition between CK and JJ during the warming phase, while the bacterial community compositions of CK and JJ were similar during the thermophilic and cooling phases. This suggests that the inoculation of the microbial agent can change the growth state of the bacterial community to a certain extent. The CK8 and RC24 samples were relatively close, indicating that the bacterial community composition of the sample from day 8 of the warming phase of CK was similar to that of the sample from the cooling phase of the RC fermentation at 24 h. This suggests that inoculation accelerated the heating rate of the RC treatment group, promoted fermentation and humification, and shortened the fermentation cycle.

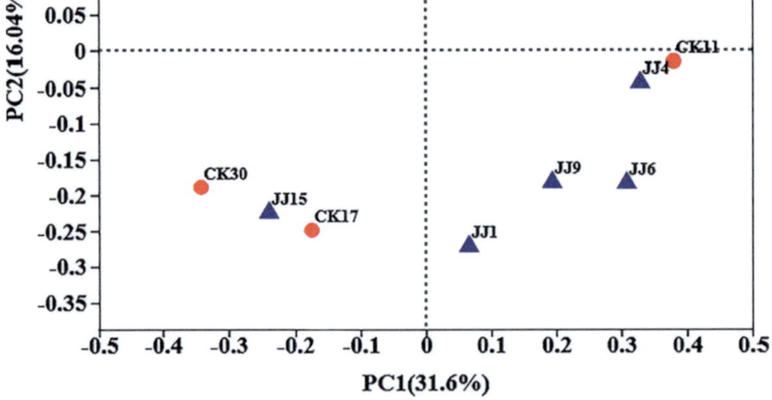

Fig. 2.24 PCoA principal coordinate analysis of bacterial communities

Response Relationship between Functional Bacteria and Environmental Factors: The dynamic succession of the microbial community during the fermentation process reflects its ability to degrade the fermentation material. In this process, the changes in the microbial community largely depend on the composition of the raw materials, environmental conditions, and the interactions between these factors. Therefore, Canonical Correspondence Analysis (CCA) was used to further investigate the relationship between environmental factors and the succession of the functional bacterial community during the fermentation of food waste, in order to determine the main environmental factors controlling the distribution of the functional bacterial community structure. Table 2.4 shows that 0.65 and 0.52 are the eigenvalues of the first two axes. The correlation coefficients of the first and second species and environmental factor ranking axes are 0.956 and 0.926, respectively, which together explain 63.55% of the total variance of the species-environment relationship. The red vector lines represent different physicochemical parameters. The length of the vector line of the arrow can represent the correlation between different physicochemical factors and the bacterial community. The longer the vector line, the greater the correlation, and vice versa. The contribution rates of each parameter to bacterial growth were TOC (Total Organic Carbon 29%), OM (Organic Matter 25.10%), and T (Temperature 24.40%), which were significantly higher than EC (Electrical Conductivity 10.80%) and pH (10.70%). This indicates that EC and pH had a weak influence on the distribution of the functional colony structure, while TOC, OM, and T ($p < 0.05$) were significant driving factors for the differences at the bacterial genus level during the fermentation of food waste.

As shown in Fig. 2.25, in the second quadrant, the species *Pseudomonas* from the warming phase of CK, JJ, and RC and the cooling and maturation phase of CK was positively correlated with TOC, OM, and EC, and negatively correlated with T and pH. *Pseudomonas* has a carbon fixation function. In the third quadrant, the species *Bacillus* and *unclassified_f__Rhizobiaceae* from the warming phase of JJ and the cooling phase of CK and JJ were only positively correlated with pH and negatively correlated with all other environmental factors. This indicates that the above environmental factors have a significant impact on the succession of the bacterial community, and improving the environmental impact factors provides a potential possibility for increasing the abundance and diversity of core functional microorganisms.

Table 2.4 CCA ordination eigenvalues, correlation coefficients of species and environmental factor ordination axes

Axis	Axis 1	Axis 2	Axis 3	Axis 4
Eigenvalue	0.6511	0.5053	0.4738	0.1236
Species-environment correlation	0.9558	0.9255	0.8409	0.7851
Cumulative percentage of species (%)	18.74	33.28	46.91	50.47
Cumulative percentage of species-environment correlation (%)	35.78	63.55	89.59	96.38

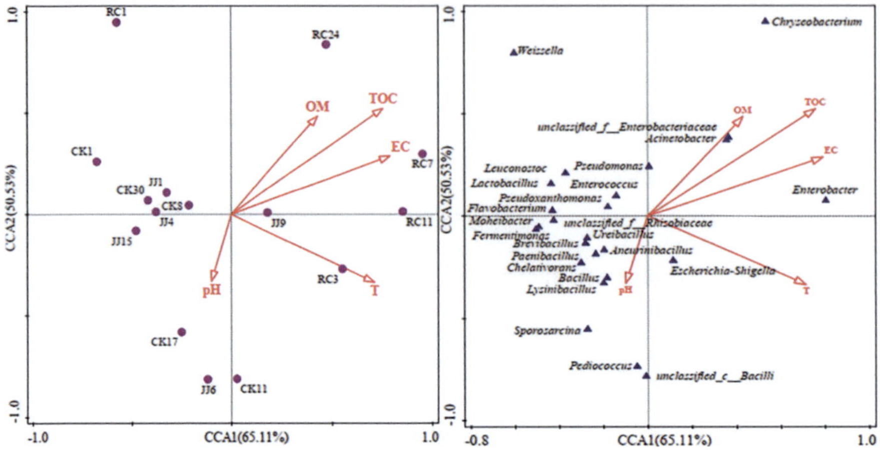

Fig. 2.25 CCA analysis of bacterial genus level and environmental factors

Response Relationship between Functional Bacteria and Organic Carbon Components First, Detrended Correspondence Analysis (DCA) was performed on the species data. If the gradient length of the first axis (Lengths of gradient) is < 3, Redundancy Analysis (RDA) is used. If it is > 4, Canonical Correspondence Analysis (CCA) is used. The results showed that the largest gradient length was 2.30, so CCA analysis was chosen. In the figure, the red arrows represent organic carbon components, and the purple triangles represent functional bacterial genera. As shown in the CCA analysis results in Fig. 2.26, the contribution rate of the organic carbon components was ranked as II (38.00%) > III (26.20%) > V (20.90%) > I (15.00%) > IV (0.70%). Components I and IV were negatively correlated with functional bacteria 4 (*Pseudomonas*), 5 (*Bacillus*), and 18 (*unclassified_f__Rhizobiaceae*). Component II was significantly negatively correlated ($p < 0.05$) with the functional bacteria *Pseudomonas*, *Bacillus*, and *unclassified_f__Rhizobiaceae*. Components III and V were positively correlated with Pseudomonas, Bacillus, and unclassified_ f__Rhizobiaceae. The *Bacillus* genus and the *unclassified_f__Rhizobiaceae* family have nitrogen fixation functions, and Bacillus can significantly increase the content of humus carbon in fermentation products, which helps to improve soil quality and crop growth, while *Pseudomonas* has a carbon fixation function. This indicates that the *Pseudomonas* genus, *Bacillus* genus, and *unclassified_f__Rhizobiaceae* family are very likely related to the decomposition of protein-like substances and soluble microbial byproducts and the synthesis of humic-like substances. At the same time, small-molecule organic carbon components (protein-like substances) have the potential to increase the abundance of functional bacteria. Combined with the analysis of Fig. 2.26, it was found that carbon-fixing bacteria (Pseudomonas) were significantly positively correlated with TOC and OM, and negatively correlated with T and pH. At the same time, they were negatively correlated with small-molecule organic carbon components (protein-like substances, soluble microbial byproducts).

Nitrogen-fixing bacteria were positively correlated with pH and negatively correlated with all other environmental factors. The above results indicate that changes in environmental factors in the material and the transformation of organic carbon components are all correlated with the succession of the microbial community. For example, temperature changes affect the metabolic activity of bacteria, which in turn affects the speed at which they utilize organic matter. That is, when the temperature is high, the decomposition rate of organic matter is accelerated, the easily degradable organic carbon components in the material decrease, the nutrient sources decrease, which in turn affects microbial metabolic activities; when the temperature is low, the metabolic activities of microorganisms are weakened, the rate of transformation of organic matter is slow, which affects the humification process. Therefore, fermentation is accomplished through the interaction of functional bacteria, organic carbon components, and environmental factors.

Through the above analysis, the following conclusions were obtained:

(1) Bacterial community diversity analysis showed that in the fermentation products of CK and JJ, the abundance of *Bacillus* and *unclassified_f__Rhizobiaceae*, which have nitrogen fixation functions, was high. In the product of RC, the

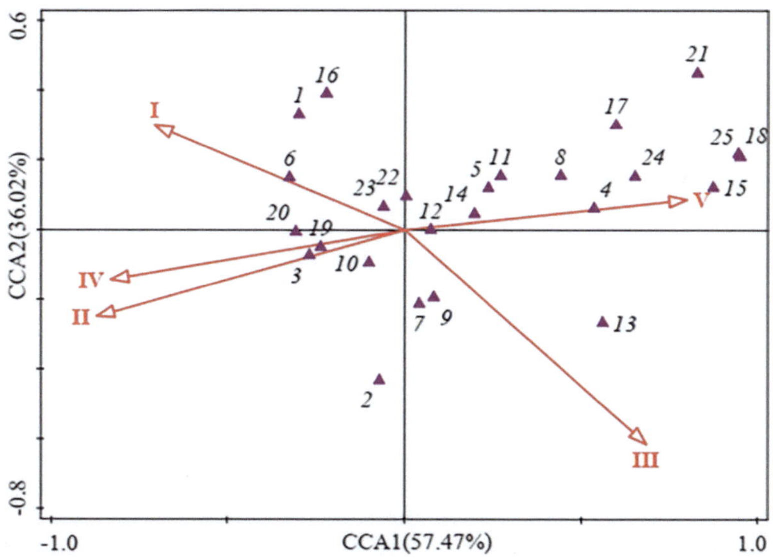

Fig. 2.26 CCA analysis of functional bacterial genera and organic carbon components. 1 *Lactobacillus*, 2 *Enterobacter*, 3 *Pediococcus*, 4 *Pseudomonas*, 5 *Bacillus*, 6 *Weissella*, 7 *Acinetobacter*, 8 *Ureibacillus*, 9 *unclassified_F__Enterobacteriaceae*, 10 *Brevibacillus*, 11 *Lysinibacillus*, 12 *Aneurinibacillus*, 13 *Chryseobacterium*, 14 *Paenibacillus*, 15 *Moheibacter*, 16 *Leuconostoc*, 17 *Chelativorans*, 18 *unclassified_F__Rhizobiaceae*, 19 *Escherichia-Shigella*, 20 *unclassified_C__Bacilli*, 21 *Sporosarcina*, 22 *Pseudoxanthomonas*, 23 *Enterococcus*, 24 *Flavobacterium*, 25 *Fermentimonas*, I. Tyrosine-like substance, II. Tryptophan-like substance, III. Fulvic-like acid substance, IV. Soluble microbial byproduct, V. Humic-like acid substance

abundance of *Pseudomonadaceae*, whose main function is carbon fixation, was high.

(2) The results of the response analysis of functional bacteria and environmental factors showed that TOC, OM, and T were the main controlling factors for the differences at the bacterial genus level. They were significantly negatively correlated ($p < 0.05$) with the nitrogen-fixing *Bacillus* and *unclassified_f__Rhizobiaceae*. The carbon-fixing *Pseudomonadaceae* was significantly positively correlated ($p < 0.05$) with TOC and OM, and significantly negatively correlated ($p < 0.05$) with T. This indicates that environmental factors have a significant impact on the succession of the functional microbial community, and environmental factors can be appropriately adjusted to regulate the succession of the microbial community.

(3) The response analysis of functional bacteria and organic carbon components showed that functional bacteria were negatively correlated with small-molecule organic carbon components like protein-like substances and soluble microbial byproducts, and positively correlated with large-molecule organic carbon components like humic-like substances. This indicates that functional bacteria may promote the decomposition of small-molecule organic carbon components and the synthesis of humic-like substances. Combined with the response analysis of functional bacteria and environmental factors, it is shown that functional bacteria, organic carbon components, and environmental factors interact and influence each other to complete the fermentation.

2.4.2 Fungal Community Succession During Fermentation

2.4.2.1 Analysis of Fungal Community Diversity

The analysis process for fungal community diversity was similar to that for the bacterial community in the previous section. The optimized sequences for CK, JJ, and RC were 328, 221, 351, 651, and 324, 436, respectively, corresponding to 176, 127, and 320 OTUs (Operational Taxonomic Units). A Venn diagram analysis (Fig. 2.27) showed a total of 414 fungal OTUs across the three fermentation methods. Among these, 46 fungal OTUs were unique to CK, 28 were unique to JJ, and 194 were unique to RC. The three fermentation processes shared 62 OTUs. CK and JJ shared 20 fungal OTUs, CK and RC shared 47 fungal OTUs, and JJ and RC shared 17 fungal OTUs. This indicates that RC had the highest number of fungal OTUs and differed significantly from CK and JJ, while the fungal communities of CK and JJ were more similar.

As shown in Table 2.5, a comparison of CK diversity using the Shannon and Simpson indices revealed that the Shannon index showed an overall increasing trend, while the Simpson index showed a decreasing trend. This suggests that the diversity

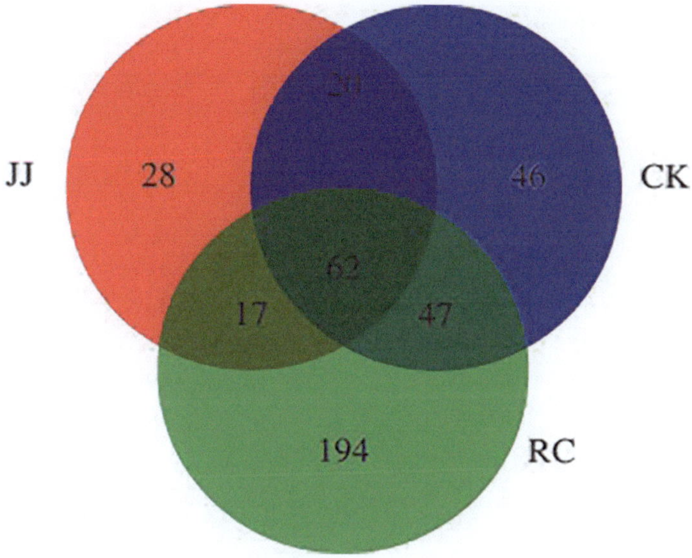

Fig. 2.27 Venn diagram at the fungal OTU level

of the fungal community in the CK samples increased as fermentation progressed. The ACE and Chao1 indices showed an overall upward trend, but both fluctuated during the fermentation process, indicating that the community richness varied at different stages.

The changes in fungal diversity indices in JJ were similar to those in CK. However, at the end of fermentation, the ACE and Chao1 index values were higher than in CK, indicating that the abundance of the fungal community in JJ was higher than in CK at the final stage of fermentation. In RC, the Shannon, ACE, and Chao1 indices showed a continuous downward trend, while the Simpson index increased. This indicates that both the diversity and abundance of the fungal community in the RC samples decreased as the fermentation process continued. This may be because the temperature was maintained at 71 °C from the 5th to the 7th hour of fermentation; this excessively high temperature inactivated most fungi. Additionally, the pH of the material remained around 5, creating a weakly acidic microenvironment unsuitable for microbial survival and reproduction. The order of the Shannon index values for the three fermentation processes was CK > JJ > RC. The order for the ACE and Chao1 indices was JJ > CK > RC. This shows that the CK group had higher fungal community diversity, the RC group had the lowest, and the JJ group had a higher abundance of specific fungal communities. The effective sequence Coverage for all three groups of fermented samples exceeded 99%, indicating that they could represent the true situation of the fungal communities in the samples.

Table 2.5 Fungal diversity indices

Sample No.	Shannon	Simpson	ACE	Chaol	Coverage
CK1	1.52	0.33	78.63	69.00	0.9997
CK8	0.21	0.92	52.95	40.00	0.9998
CK11	0.05	0.99	16.24	13.00	0.9999
CK17	1.42	0.48	112.16	111.75	0.9998
CK30	2.58	0.13	100.45	107.50	0.9998
JJ1	2.35	0.14	81.54	78.11	0.9997
JJ4	1.23	0.51	58.52	49.00	0.9998
JJ6	0.22	0.94	33.64	33.33	0.9999
JJ9	1.79	0.24	80.56	59.60	0.9997
JJ15	2.49	0.14	114.70	110.50	0.9996
RC1	3.70	0.06	219.93	220.50	0.9998
RC3	2.00	0.36	137.64	137.67	0.9999
RC7	2.59	0.11	44.81	45.00	0.9999
RC11	2.20	0.23	85.19	85.33	0.9999
RC24	1.12	0.63	58.74	58.33	1.0000

2.4.2.2 Composition and Dynamic Succession of Fungal Communities

Changes at the Fungal Community Phylum Level: During aerobic fermentation, fungi not only have a strong ability to degrade recalcitrant organic matter such as cellulose, but their mycelia also have a mechanical penetration effect that helps break down the material, promoting its degradation. Fungi and bacteria work synergistically to degrade lignin, with fungi playing a major role in its decomposition. The ability of fungi to decompose lignocellulose can directly affect the rate and quality of fermentation. Figure 2.28 shows the relative abundance at the phylum level for the three treatment groups. Species with a relative abundance of < 0.01% were classified as "Others". In the initial stage of fermentation, the main dominant phylum in the CK group was *Ascomycota*, with an abundance as high as 99%. As the fermentation time was extended, the temperature decreased and the types of organic matter changed, leading to a change in its abundance. By the end of fermentation, its abundance had decreased by 64.92%, and *Mucoromycota* (21.48%) and *Basidiomycota* (38.08%) emerged as new dominant phyla.

The abundance of *Ascomycota* remained high throughout the JJ fermentation period, reaching 84.47% at the end of fermentation. This is because *Ascomycota* can secrete a variety of enzymes that degrade cellulose, hemicellulose, and lignin, allowing it to utilize a rich nutritional substrate and fully degrade organic matter. The abundances of *Basidiomycota* and *Mucoromycota* were 14.52% and 0.14%, respectively. In the initial stage of fermentation in the RC group, the abundance ratio of

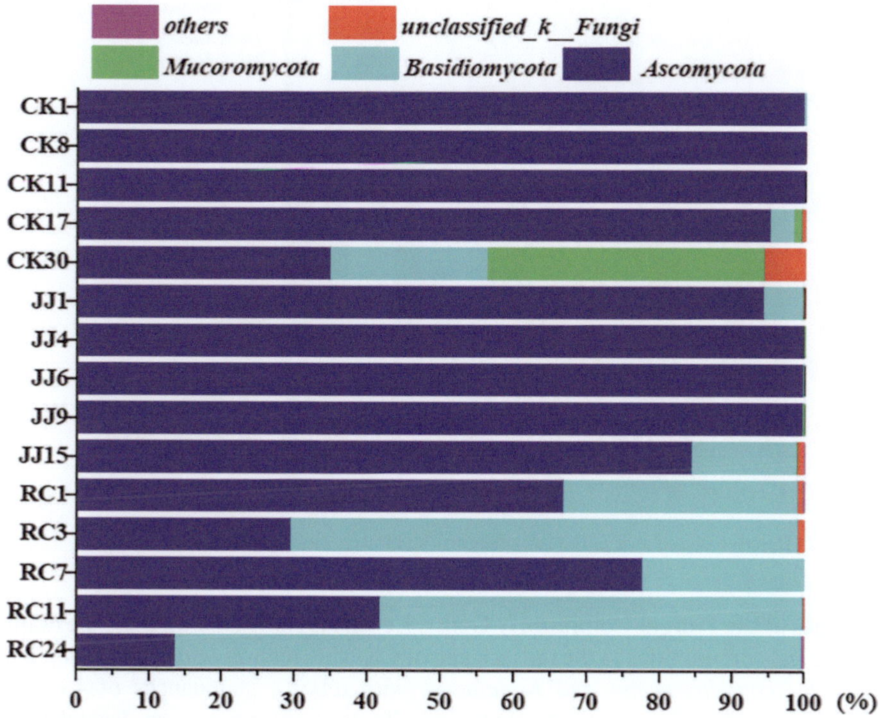

Fig. 2.28 Relative abundance at the fungal phylum level

Ascomycota to *Basidiomycota* was 2:1, making them the main dominant phyla. *Basidiomycota* was present in all stages of fermentation, and by the end of fermentation, its abundance reached 86.15%, evolving into the absolute dominant phylum in the material. In contrast, *Ascomycota* was almost absent during the fermentation process. *Basidiomycota* plays an important role in soil nutrient acquisition, decomposition, and carbon sequestration. Furthermore, *Basidiomycota* can increase the solubility of poorly soluble phosphorus, causing the EC (electrical conductivity) value of RC to rise continuously. Research has shown that fungi have poor heat tolerance. When the temperature exceeds 60 °C, some fungi die or become dormant. Therefore, the fungal diversity in JJ and RC at the end of fermentation was lower than in CK. Some studies have found that each microbial group acts on the degradation of its specific organic matter. As the substrate components of the material change, different fungal communities appear in succession. The continuous changes in fungal species and abundance constantly alter the microenvironment within the material. Competition and synergy among microorganisms gradually lead to the succession of a complex and stable community structure, promoting the further decomposition and transformation of organic matter in food waste until maturation.

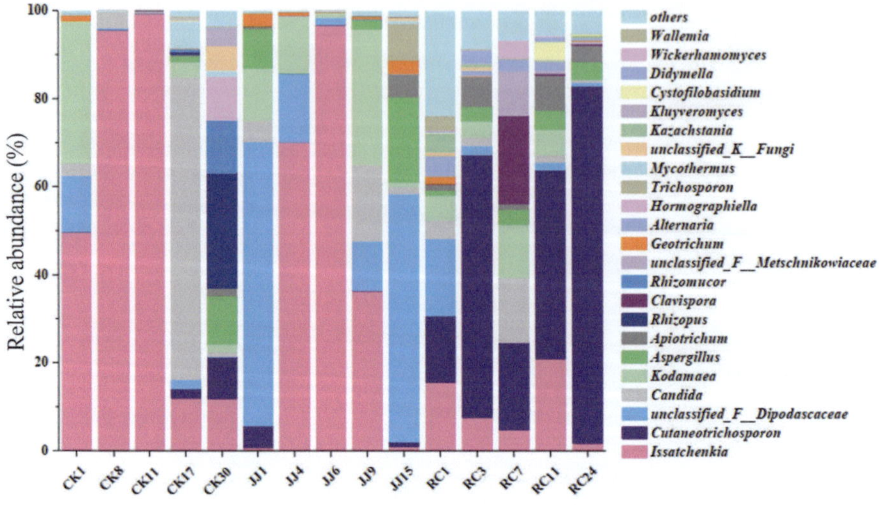

Fig. 2.29 Relative abundance at the fungal genus level

Changes at the Fungal Community Genus Level: As shown in Fig. 2.29, the dominant fungal communities in the initial stage of CK were mainly *Issatchenkia, unclassified_F__Dipodascaceae*, and *Kodamaea*, with relative abundances of 49.75%, 12.73%, and 32.22%, respectively. In the middle stage of fermentation, *Issatchenkia* became the main dominant species in the material, with its relative abundance exceeding 95%. By the late stage of fermentation, it dropped to 11% but still maintained its status as a dominant species. Therefore, *Issatchenkia* played a crucial role in the CK fermentation. Due to external inoculation, the relative abundances of *unclassified_F__Dipodascaceae*, *Candida*, and *Kodamaea* in JJ and RC were higher than in CK during the middle stage of fermentation. *Candida* can decompose hemicellulose in the material, but its heat tolerance is poor, so its abundance decreased in the later stage of fermentation, and the fungal community underwent succession.

In the final stage of fermentation in CK, the dominant fungal communities expanded to include *Cutaneotrichosporon*, *Aspergillus*, *Rhizopus*, *Rhizomucor*, *Hormographiella*, and *Kluyveromyces*, with relative abundances of 9.44%, 10.96%, 26.05%, 12.03%, 9.88%, and 4.20%, respectively. Among these, the mycelia of *Aspergillus* and *Rhizopus* have a strong mechanical penetration effect and work together to degrade recalcitrant organic matter such as cellulose and pectin. Combined with three-dimensional fluorescence data, it was found that the content of large-molecule humic-like substances was over 80% at the end of fermentation. This may be because the further degradation of cellulosic materials by fungi such as *Aspergillus* and *Rhizopus* in the final stage of fermentation provided a large number of precursor substances for the synthesis of humic-like substances, thereby increasing their content. Hydroxyl groups play an important role in the decomposition of lignin by fungi. A higher content of hydroxyl functional groups can promote the degradation of lignin by fungi, resulting in a higher content of humic-like substances in the final product of CK fermentation compared to the other two groups. The dominant fungal groups in JJ at the end of fermentation were *unclassified_F__Dipodascaceae* (56.40%), *Aspergillus* (19.25%), *Apiotrichum* (5.19%), and *Trichosporon* (8.23%). In RC, the dominant genera at the end of fermentation included *Issatchenkia, Cutaneotrichosporon, unclassified_F__Dipodascaceae, Aspergillus*, and *Apiotrichum.*

Principal Coordinate Difference Analysis of Fungal Communities: Based on the Unweighted UniFrac algorithm (which only considers differences in species categories among samples) and the Weighted UniFrac algorithm (which considers both differences in species categories and abundance among samples), a principal coordinate analysis (PCoA) was performed on the samples (Fig. 2.30). The PCoA analysis revealed the differences in the fungal communities of food waste during the fermentation process under different biological treatment methods, with PCA1 and PCA2 contributing 23.23% and 19.02%, respectively. The relative positions between the sample points show the differences between the communities; the farther the distance, the greater the difference in community composition, and vice versa. The fungal sample points for CK and JJ shifted at different fermentation periods. The JJ samples from day 4 and day 6 were close to the CK samples from day 8 and day 11, indicating that the fungal community composition of JJ during its high-temperature phase was similar to the fungal community structure of CK during its high-temperature phase. This suggests that the difference between the JJ and CK communities was small during the high-temperature phase. However, the samples from the two groups were far apart during other fermentation stages, indicating greater differences between the CK and JJ communities at other times. The RC7 sample was close to the CK1 sample, indicating that the community structure of RC during its high-temperature phase was similar to the community structure of CK during its temperature-rise phase.

Relationship Between Functional Fungi and Environmental Factors: Canonical Correspondence Analysis (CCA) was used to further investigate the relationship between the fungal community structure at the genus level and environmental factors, and to identify the key environmental elements affecting the succession of functional fungal communities. As shown in Fig. 2.31, TOC ($p < 0.01$) was a highly significant environmental factor affecting the fungal community structure, and OM ($p < 0.05$) was a significant environmental factor. It has been reported that or ganic carbon components are essential nutrients for community growth. Cutaneotrichosporon was positively correlated with T, TOC, OM, and EC, and negatively correlated with pH. It was also positively correlated with the temperature-rise and high-temperature phases of the RC treatment group. In the second quadrant, the species *Rhizopus* and *unclassified_F__Dipodascaceae* from the cooling-down phase of the CK and JJ groups were positively correlated with pH and negatively correlated with T, TOC, OM, and EC. In the fermentation product of CK, *Rhizopus* was one of the main dominant species, while in JJ, it was *unclassified_F__Dipodascaceae.* The main

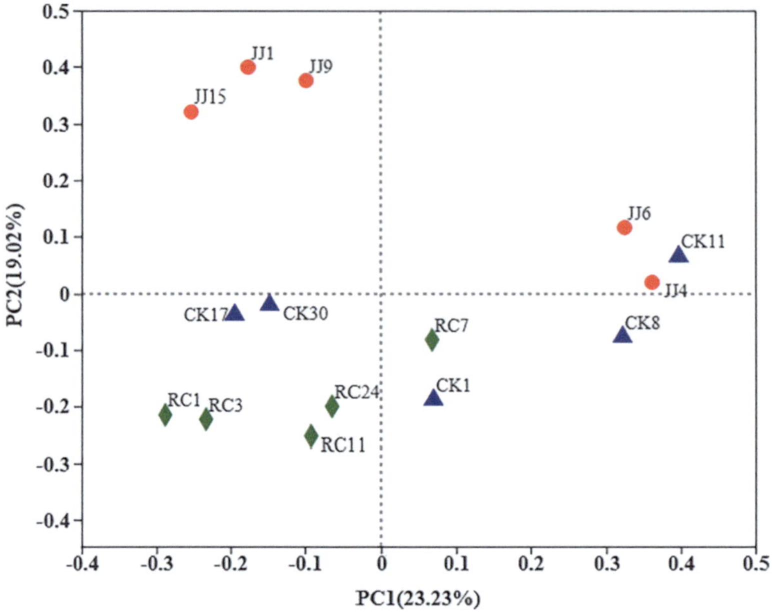

Fig. 2.30 PCoA principal coordinate analysis of fungal communities

function of these two fungi is phosphate solubilization. In RC, the main dominant genus was *Cutaneotrichosporon.*

Relationship Between Functional Fungi and Organic Carbon Components: Detrended Correspondence Analysis (DCA) was performed on the organic carbon component and fungal community data. The result showed that the gradient value of the first axis was 2.9, which is less than 3. Therefore, Redundancy Analysis (RDA) was chosen to reveal the response relationship between organic carbon components and functional fungal communities. The length and angle of the arrows represent the importance and correlation of the variables. An angle of less than 90° indicates a positive correlation, with a smaller angle indicating a stronger positive correlation. An angle greater than 90° indicates a negative correlation, with a larger angle indicating a stronger negative correlation. The eigenvalues of the first two ordination axes were 0.2748 and 0.0929, explaining 36.77% of the variance. The red vector lines represent the organic carbon components, and the blue vector lines represent the functional fungal genera. Figure 2.32 shows the relationship between the organic carbon components and the functional fungal genera. The contribution rate of organic carbon components affecting the diversity and abundance of the fungal community was III (52.10%) > IV (26%) > II (18%) > I (3.9%) > V (<0.1%). Therefore, III (fulvic acid-like substances) is the main organic carbon component affecting the structure of the functional fungal community. Organic carbon components I, II, and IV were negatively correlated with 2 (*Cutaneotrichosporon*), 3 (*unclassified_F__Dipodascaceae*), and 8 (*Rhizopus*). The collinearity

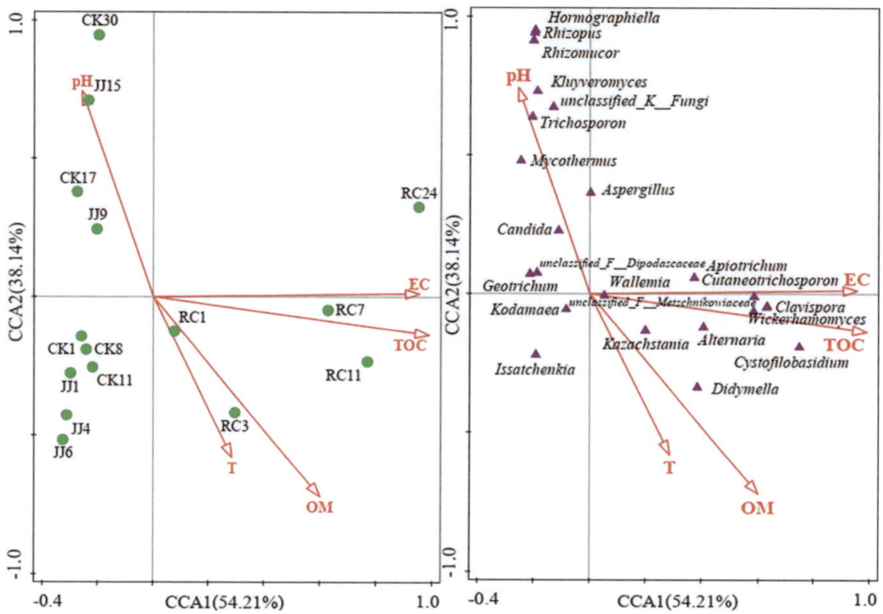

Fig. 2.31 CCA analysis of fungal genus level and environmental factors

of II and IV indicates a strong correlation between them. Component III was significantly positively correlated with the fungal genera *Cutaneotrichosporon*, *unclassified_F__Dipodascaceae*, and *Rhizopus* ($p < 0.05$). Component V was negatively correlated with *Cutaneotrichosporon* and positively correlated with *unclassified_F__Dipodascaceae* and *Rhizopus*. *Cutaneotrichosporon* belongs to the phylum *Basidiomycota*. *Basidiomycota* has a carbon sequestration function and was the dominant genus in RC. This functional genus was negatively correlated with pH, and fulvic acid-like substances were also significantly negatively correlated with pH. Furthermore, *Cutaneotrichosporon* was significantly positively correlated with fulvic acid-like substances. This indicates that regulating pH affects the abundance of functional bacteria in the material, which in turn affects the transformation of organic carbon components. In JJ, *unclassified_F__Dipodascaceae* was dominant. It can improve the utilization of phosphate in the soil, converting it into a form that is easily absorbed and utilized. On the 17th day of CK fermentation, *Candida* became its absolute dominant community. *Candida* can directly promote plant growth, which is why the seed germination rate for CK was higher. In the final product of CK, the genus *Rhizopus* had the highest abundance. Its main function is phosphate solubilization. It was negatively correlated with small-molecule protein-like substances and soluble microbial by-products, and positively correlated with the content of humic-like substances. It was also positively correlated with pH and negatively correlated with TOC, OM, T, and EC. The above results indicate that organic carbon components, environmental factors, and functional fungal genera interact to complete the fermentation process.

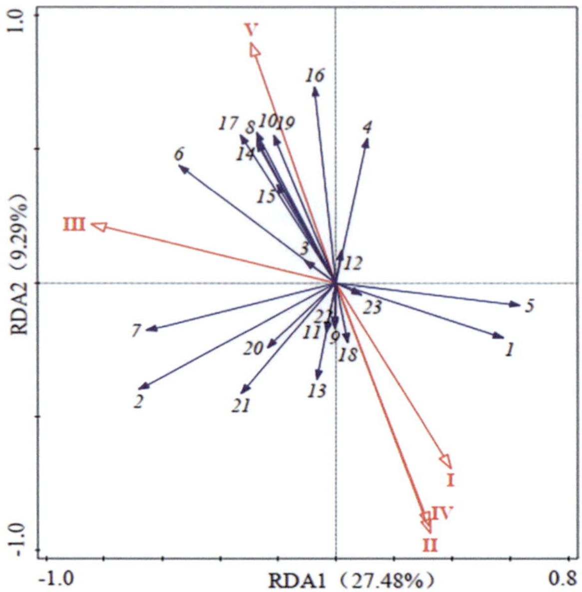

Fig. 2.32 RDA analysis of functional fungal genera and organic carbon components. 1 *Issatchenkia*, 2 *Cutaneotrichosporon*, 3 *unclassified_F__Dipodascaceae*, 4 *Candida*, 5 *Kodamaea*, 6 *Aspergillus*, 7 *Apiotrichum*, 8 *Rhizopus*, 9 *Clavispora*, 10 *Rhizomucor*, 11 *unclassified_F__Metschnikowiaceae*, 12 *Geotrichum*, 13 *Alternaria*, 14 *Hormographiella*, 15 *Trichosporon*, 16 *Mycothermus*, 17 *unclassified_K__Fungi*, 18 *Kazachstania*, 19 *Kluyveromyces*, 20 *Cystofilobasidium*, 21 *Didymella*, 22 *Wickerhamomyces*, 23 *Wallemia*., I. Tyrosine-like substances, II. Tryptophan-like substances, III. Fulvic acid-like substances, IV. Soluble microbial by-products, V. Humic acid-like substances

Through the above analysis, the following conclusions were obtained:

(1) Fungal community diversity analysis showed that at the end of fermentation, the relative abundance of *Ascomycota* was high in the CK and JJ fermentation products. In the RC fermentation product, *Basidiomycota* was one of the dominant phyla, with an abundance of 86.15%, and it has a carbon sequestration function.
(2) The analysis of the response of functional fungi to environmental factors showed that TOC ($p < 0.01$) and OM ($p < 0.05$) were the main controlling factors affecting the succession of functional fungal communities. The dominant *Rhizopus* in CK and the dominant *unclassified_F__Dipodascaceae* in JJ were positively correlated with pH and negatively correlated with TOC, OM, T, and EC. Both fungi have a phosphate solubilizing function. The abundance of the genus *Cutaneotrichosporon* was high in RC and was negatively correlated with pH and positively correlated with TOC, OM, T, and EC.
(3) Fulvic acid-like substances were the key organic carbon component affecting the succession of functional fungal communities ($p < 0.05$) and were significantly positively correlated with *Rhizopus* in CK, *unclassified_F__Dipodascaceae* in

JJ, and *Cutaneotrichosporon* in RC. The abundances of *Rhizopus* and *unclassified_F__Dipodascaceae*, which have phosphate solubilizing functions, were high in CK and JJ. Both were negatively correlated with small-molecule protein-like substances and soluble microbial by-products, and positively correlated with the content of humic-like substances.

2.5 Summary

Soil conditioners with distinct organic carbon components were produced by fermenting food waste under different biological treatment regimes. The transformation of organic carbon components in these conditioners were characterized using spectroscopic techniques. High-throughput sequencing was employed to analyze the succession patterns of microbial community structures, investigate key environmental factors influencing organic carbon component transformation, and elucidate the response relationships between organic carbon components and functional microorganisms. The main conclusions are as follows:

(1) The ratios of macromolecular (MW > 5000 Da) to low-molecular-weight organic carbon (MW < 500 Da) in CK, JJ, and RC were 4.26:1, 2.72:1, and 1.12:1, respectively. This indicates that the traditional conditioner CK primarily comprised humic carbon as its main organic carbon component. The JJ group exhibited a ratio close to 3:1, correlating with its higher germination rate and superior germination-promoting effect. For RC, the content of macromolecular humic carbon was comparable to that of low-molecular-weight organic carbon components, providing a foundation for synergistic regulation between these fractions.

(2) Three-dimensional excitation-emission matrix (EEM) fluorescence spectroscopy data revealed the highest degree of humification in CK, followed by JJ, and then RC. Fourier transform infrared (FTIR) spectroscopy indicated a higher carboxyl (COO^-) content in JJ compared to CK and RC, signifying elevated levels of carboxylic acid substances in JJ. RC exhibited the highest hydroxyl (–OH) content among all treatments, suggesting greater abundance of low-molecular-weight alcohols and phenols. Ultraviolet–visible (UV–Vis) spectroscopy data showed that the magnitudes of three characteristic values followed the order CK > JJ > RC, further confirming CK's highest humification degree and largest molecular weight, followed by JJ, and then RC.

(3) Correlation analysis demonstrated significant or highly significant positive correlations between pH and $SUVA_{280}$, indicating a substantial influence of pH on molecular weight. Increasing pH may promote the formation of high-molecular-weight organic carbon components. In CK and JJ treatments, organic matter (OM) and total organic carbon (TOC) showed significant or highly significant positive correlations with protein-like substances and highly significant negative correlations with humic-like substances. This suggests OM and TOC

are primary factors influencing the formation of various organic carbon components; increasing OM content may promote the formation of low-molecular-weight protein-like substances. The formation of humic-like substances in RC was primarily influenced by temperature and pH.

(4) Microbial diversity analysis showed higher abundance of nitrogen-fixing functional bacteria, namely *Bacillus* (genus) and unclassified_f__*Rhizobiaceae* (family), in the fermentation products of CK and JJ. The RC product exhibited higher abundance of *Pseudomonadaceae* (family), known for carbon sequestration. Ascomycota was the dominant fungal phylum in CK and JJ products. In contrast, Basidiomycota was the dominant phylum (86.15% abundance) in the RC fermentation product, representing a 4- to fivefold increase compared to CK and JJ treatments. Basidiomycota primarily functions in carbon fixation. These results indicate superior nitrogen fixation effects in CK and JJ products, while the RC product demonstrates stronger carbon sequestration capacity.

(5) Analysis of the response relationships among environmental factors, organic carbon component transformation, and functional microbial community succession revealed that TOC, OM, and temperature (T) were the dominant factors governing the succession of functional bacterial communities (*Bacillus*, unclassified_f__*Rhizobiaceae*, and *Pseudomonadaceae*). TOC and OM were key factors affecting the abundance and diversity of functional fungal communities (*Rhizopus* and *Basidiomycota*). Functional bacteria exhibited negative correlations with low-molecular-weight protein-like substances and soluble microbial byproducts, but positive correlations with macromolecular humic-like substances. These findings indicate that distinct physicochemical factors create different survival environments for microorganisms, influencing microbial community succession and the transformation of organic carbon components. Conversely, organic carbon components provide carbon and nutrient sources for microbial growth and reproduction, thereby affecting microbial community succession. Consequently, appropriately regulating environmental factors can steer microbial community succession and organic carbon component transformation within the material.

References

Awasthi SK, Wong JWC, Li J (2018) Evaluation of microbial dynamics during post-consumption food waste composting. Biores Technol 251:181–188

Hou J, Li M, Xi B (2017) Short-duration hydrothermal fermentation of food waste: preparation of soil conditioner for amending organic-matter-impoverished arable soils. Environ Sci Pollut Res 24(26):21283–21297

Kong L, Dong W, Yang T, Ma X, An N, Zhou X, Zhu Y (2019) The effect of humic acid in improving and enriching soil. Humic Acid 1:6–12

Li L (2014) Preparation of composite microbial inoculation and its promotion on chicken manure composting. Tsing Hua University, Beijing

Song C, Li M, Wei Z (2016) Effect of inoculation with multiple composite microorganisms on characteristics of humic fractions and bacterial community structure during biogas residue and livestock manure co-composting. J Chem Technol Biotechnol 91(1):155–164

Xi B, Meng W, Liu H, Huang G, Zeng G, Wang Q (2003) The variation of inoculation complex microbial community in three stages MSW composting process controlled by temperature. Environ Sci 02:152–155

Xi B, He X, Dang Q (2015) Effect of multi-stage inoculation on the bacterial and fungal community structure during organic municipal solid wastes composting. Biores Technol 196:399–405

Yue D, Wang L, Qiao L, Li D, Zan L (2018) Screening of high-efficiency cellulose-degrading strains and application effects of their compound microbial agents in organic composting. Jiangsu Agric Sci 46(17):273–276

Zhu L, Wei Z, Yang T et al (2020) Core microorganisms promote the transformation of DOM fractions with different molecular weights to improve the stability during composting. Biores Technol 299:122575

Chapter 3
Preparation and Properties of a Synergistic and High-Efficiency Soil Conditioner

Abstract Food waste contains a large amount of easily degradable organic matter, such as sugars, starches, fats, and proteins, as well as some nutrients that are readily utilized by plants. Preparing soil conditioners from food waste is not only an effective way to recycle and reuse food waste but can also be used to improve soil fertility. Soil conditioners prepared by traditional fermentation methods are rich in humic carbon but contain less small molecular available carbon. When applied as a base fertilizer, they have a significant effect on increasing crop yields in fertile farmland. However, for barren soil, which has low organic matter content and a dysfunctional microbial community, the application of organic fertilizer often fails to fundamentally enhance the ecological function of its own microbial community. Therefore, we propose a hypothesis: the formation of humus in barren soil can use macromolecular humic carbon as a carbon skeleton to promote the condensation of soil aggregates, while the increase in the abundance and diversity of the microbial community in barren soil can use small molecular available carbon as an energy and nutrient source. Under the synergistic regulation of small molecular available carbon and macromolecular humic carbon, barren soil can be conditioned and improved, enhancing its fertility. Based on this, this chapter uses food waste as a raw material to prepare a synergistic and high-efficiency conditioner using high-temperature hydrothermal hydrolysis equipment. With traditionally fermented organic fertilizer as a reference, the physicochemical properties, morphological characteristics, degree of humification, changes in functional groups, and microbial community structure of the synergistic and high-efficiency conditioner were characterized.

3.1 Preparation of the Soil Conditioner

Food waste contains a large amount of easily degradable organic matter, such as sugars, starches, fats, and proteins, as well as some nutrients that are readily utilized by plants (Chiew et al. 2015; Li et al. 2013). The nutrients and organic carbon lost from barren soil should be returned to the soil to maintain its fertility. Preparing soil

J. Hou et al., *Food Waste Valorization: Preparation, Properties, and Application of the Synergistic Soil Conditioners*, https://doi.org/10.1007/978-981-95-5362-4_3

conditioners from food waste is not only an effective way to recycle and reuse food waste but can also be used to improve soil fertility (Chiew et al. 2015).

Soil conditioners can be prepared through fermentation, a method that not only effectively reduces carbon loss but also saves land space (Zaher et al. 2016). During the fermentation process, the organic components in food waste can be converted into slow-release nutrients, such as stable humic-like substances, and readily available nutrients (Coelho et al. 2011; He et al. 2011; Lashermes et al. 2012). However, in most cases, traditional fermentation methods have certain technical and economic limitations. For example, due to national dietary habits, food waste in our country typically contains high levels of oil and moisture, making it difficult to degrade (Jiang et al. 2015). Furthermore, traditional fermentation methods usually require higher processing costs, including equipment maintenance and labor fees, and the fermentation cycle often lasts for about a month (Kuo and Cheng 2007). For large-scale fermentation, the traditional equipment used often requires a large footprint, which poses challenges for land management and maintenance.

Moreover, conditioners prepared by traditional fermentation methods are rich in humic carbon but contain less small molecular available carbon. When applied as a base fertilizer, they have a significant effect on increasing crop yields in fertile farmland. However, for barren soil, which has low organic matter content and a dysfunctional microbial community, the application of organic fertilizer often fails to fundamentally enhance the ecological function of its own microbial community. Therefore, we propose a hypothesis: the formation of humus in barren soil can use macromolecular humic carbon as a carbon skeleton to promote the condensation of soil aggregates, while the increase in the abundance and diversity of the microbial community in barren soil can use small molecular available carbon as an energy and nutrient source. Under the synergistic regulation of small molecular available carbon and macromolecular humic carbon, barren soil can be conditioned and improved, enhancing its fertility. Based on this, this chapter uses food waste as a raw material to prepare a synergistic and high-efficiency conditioner using high-temperature hydrothermal hydrolysis equipment.

3.1.1 High-Temperature Hydrothermal Hydrolysis and Fermentation Equipment

The high-temperature hydrothermal hydrolysis fermentation equipment consists of an automatic feeding machine (composed of a transmission device and a feeding machine), a feed inlet, a crusher (with built-in crushing blades directly driven by a motor; the crushing blades are activated by an automatic monitoring system and equipped with an emergency brake function to ensure simultaneous feeding and crushing as well as operational safety), a fermentation chamber (including a heating plate, one main stirring shaft, and two auxiliary stirring wheels), and a deodorization chamber. The high-temperature hydrothermal hydrolysis fermentation equipment is

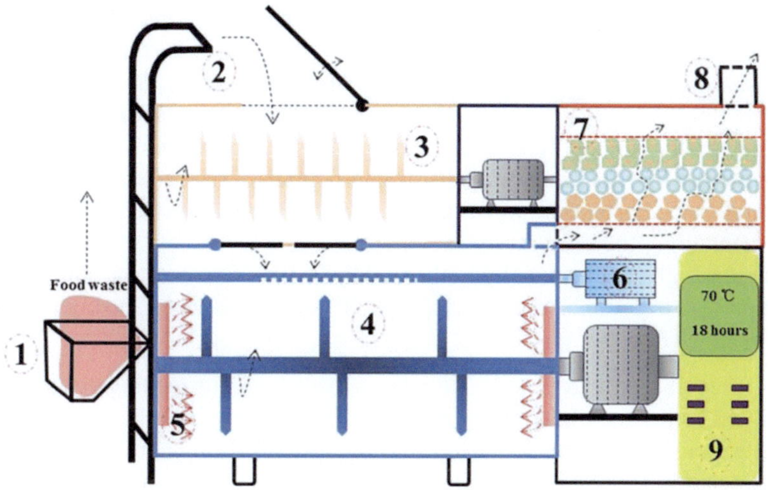

Fig. 3.1 Internal structure of the high-temperature hydrothermal hydrolysis fermentation equipment, where 1–10 represent: 1 feeding machine, 2 feed inlet, 3 crushing device, 4 fermentation reaction chamber, 5 heating, 6 fan, 7 deodorization system, 8 gas outlet, 9 control system, and 10 screw discharge outlet, respectively

configured with a programmable automatic monitoring system that can monitor the material temperature and oxygen content online and perform real-time weighing of the material. The internal structure of the equipment is shown in Fig. 3.1.

3.1.2 Preparation Method

Food waste was obtained from a centralized collection station in Beijing. Before fermentation, impurities such as plastics and large bones were sorted out. Rice husk was crushed to 1–2 cm to be used as a fermentation bulking agent. Food waste and rice husk were mixed at a wet weight ratio of 11:4 to serve as the fermentation raw material. The basic properties of the materials are shown in Table 3.1.

Table 3.1 Basic properties of materials

Material	C/N	Organic matter (%)	Moisture content (%)
Food waste	23.5 ± 0.45	87.7 ± 1.23	83.2 ± 1.22
Rice Husk	32.4 ± 0.34	89.4 ± 1.34	10.5 ± 0.37
Mixture	26.0 ± 0.42	88.2 ± 1.15	62.4 ± 1.04

The exogenously added microorganisms were a composite microbial system with the following composition: *Bacillus circulans* (10–15%), *Bacillus sphaericus* (10–15%), *Bacillus firmus* (10–15%), *Bacillus schlegelii* (5–10%), *Bacillus stearothermophilus* (10–15%), *Bacillus subtilis* (5–10%), *Candida tropicalis* (10–20%), and *Lactobacillus delbrueckii* (10–15%). The proteases, lipases, amylases, chitinases, etc., produced by these microorganisms can degrade large-molecule proteins into small-molecule substances such as polypeptides and oligosaccharides, accelerating the fermentation process.

Next, the high-temperature hydrothermal hydrolysis fermentation equipment was used to prepare traditional organic fertilizer (TF) and a synergistic and high-efficiency conditioner (SHF).

(1) TF Preparation: The mixed material was loaded into the feeding machine, transported to the feed inlet, and crushed in the crushing chamber for 30–60 min. After crushing, the material fell into the fermentation chamber. A traditional fermentation method was used with a ventilation rate of 2 mL min^{-1}. The main stirring shaft was activated every 5 days to stir for 10 min at a speed of 20 r min^{-1}. The fermentation time was 20 days. The deodorization device was on throughout the process.
(2) SHF Preparation: The mixed material was loaded into the feeding machine, transported to the feed inlet, and crushed in the crushing chamber for 30–60 min. After crushing, the material fell into the fermentation chamber, and the reaction temperature was set to 70 °C. The main stirring paddle and auxiliary stirring wheels were turned on simultaneously, with the stirring mode alternating between 20 min of forward rotation and 20 min of reverse rotation. The reaction lasted for 18 h. For the first 9 h, the material was heated to 70 °C, utilizing the high moisture content of the food waste for continuous hydrothermal hydrolysis fermentation to increase the content of small molecular available carbon. For the subsequent 9 h, the composite microbial system was inoculated at a rate of 1.25 mL kg^{-1} (1×10^8 CFU mL^{-1}) to promote the formation of humic carbon. The heating system was turned off, and the material was allowed to cool down naturally. The deodorization device was on throughout the process.

3.2 Properties of the Soil Conditioner

3.2.1 Physicochemical Characteristics

As shown in Fig. 3.2, the physicochemical indicators of TF (traditional organic fertilizer) and SHF (synergistic and high-efficiency conditioner) showed similar trends, with the exception of temperature and pH. The highest temperature for TF, 62°C, occurred on the 7th day of fermentation. The temperature during the high-temperature phase of SHF was 60.60–71.94 °C, reaching a maximum of 71.94 °C at the 14th hour of fermentation. This indicates that the continuous temperature rise of the fermentation pile was not only from the external heat source but also from the bioaugmentation effect of the exogenously added composite microbial agent and the pile's own microorganisms, which promoted the humification process of the material.

During the fermentation process, the moisture content of both TF and SHF gradually decreased. At the end of fermentation, their moisture contents were 26.55% and 22%, respectively, which were 2/5 and 1/2 of their initial values. The moisture content, through the production of metabolic water, can reflect the degradation kinetics of food waste. The compaction and porosity of the fermentation material are greatly affected by moisture content. Moisture content and stirring rate can influence the fermentation effect by adjusting the density and particle size of the reaction substrate.

During the fermentation process, the pH of TF slowly increased and then stabilized, while the pH of SHF increased at a faster rate. At the end of fermentation, the pH of SHF was 0.54 higher than that of TF. At the 12th hour of SHF fermentation, due to an ample supply of oxygen, the pH value increased from an initial value of 6.17–7.96. It has been reported that a $pH < 6.0$ will inhibit the initial stage of fermentation. However, when the pH is > 6.5 and the temperature is below 36 °C, the inhibitory effect is removed (Sundberg et al. 2004). In this experiment, the pH value was always above 6.0 during the initial stage of the rapid hydrothermal hydrolysis fermentation. As the reaction progressed, the temperature gradually increased due to external heating and the warming effect of the microbial inoculation system. Therefore, fermentation in the initial stage was not inhibited, which promoted the humification process of the material and shortened the fermentation cycle.

Electrical conductivity (EC) can reflect the total ion concentration of the fermentation material, including the concentration of soluble salts (Chen et al. 2014). In the initial stage of fermentation, the EC of both TF and SHF materials was high, possibly due to the degradation of some complex organic matter into soluble components (Wang et al. 2016). As the fermentation reaction proceeded, the EC of both TF and SHF showed minor fluctuations and began to slowly decrease in the middle stage of fermentation. The final EC values for TF and SHF were 2.36 mS cm^{-1} and 2.29 mS cm^{-1}, respectively. This value complies with the EC standard for organic fertilizer or soil conditioner applied to cultivated soil (Chen et al. 2014).

The organic matter content of both TF and SHF showed a fluctuating downward trend. The organic matter content of SHF reached a maximum of 95.56% at the 6th hour of fermentation. After fermentation, its organic matter decreased to 82%. The reason for this is that in the initial stage of the reaction, water-soluble organic matter was accumulated through stirring, but due to the mineralization of hydrophilic components, the total amount decreased (Said-Pullicino et al. 2007a).

The changes in the TC/TN (total carbon to total nitrogen) ratio for TF and SHF were similar, both showing a slow decrease followed by a stabilization trend during fermentation. At the end of fermentation, the TC/TN ratios for TF and SHF were 19.3 and 25.4, respectively. The TC/TN of SHF reached a relatively stable state after 5 h of fermentation. The reason is that in the initial stage of fermentation, some of the degradable components in the organic matter were mineralized. After 5 h

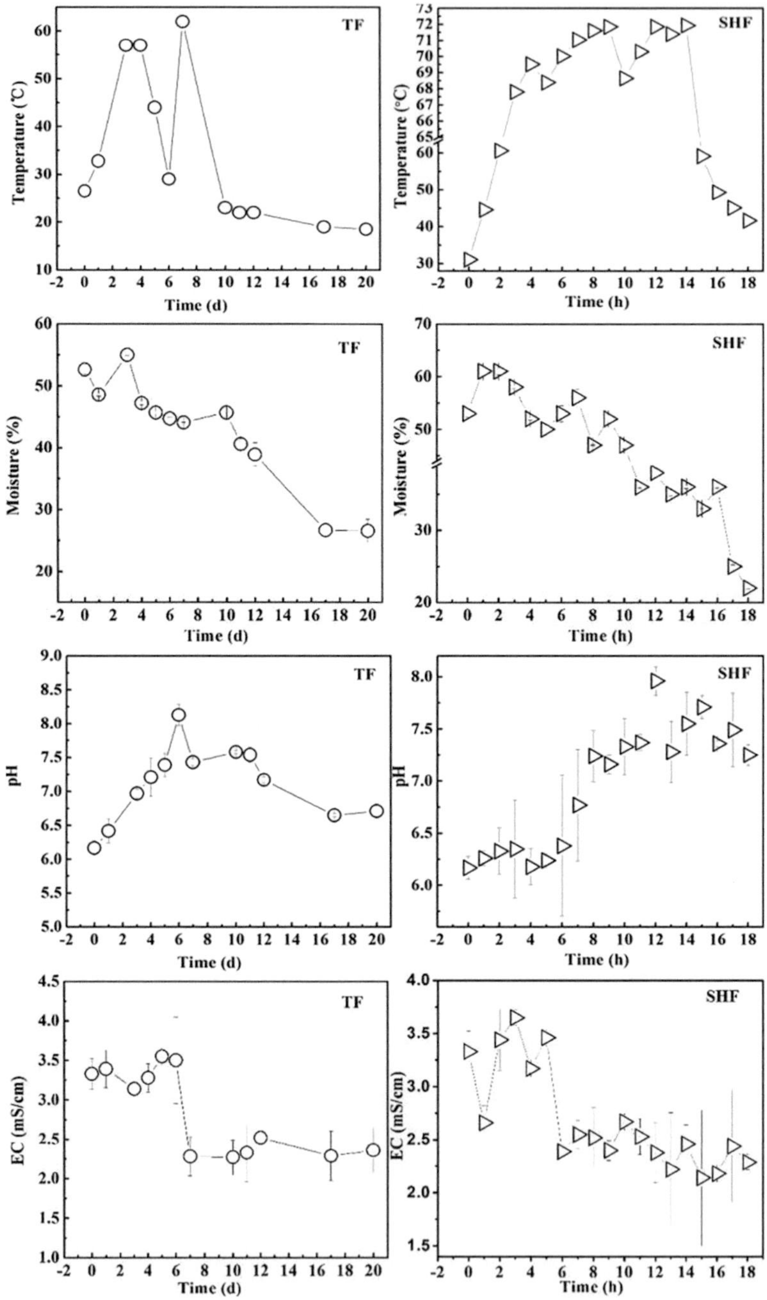

Fig. 3.2 Characterization of the physicochemical features of traditional organic fertilizer and the synergistic and high-efficiency conditioner

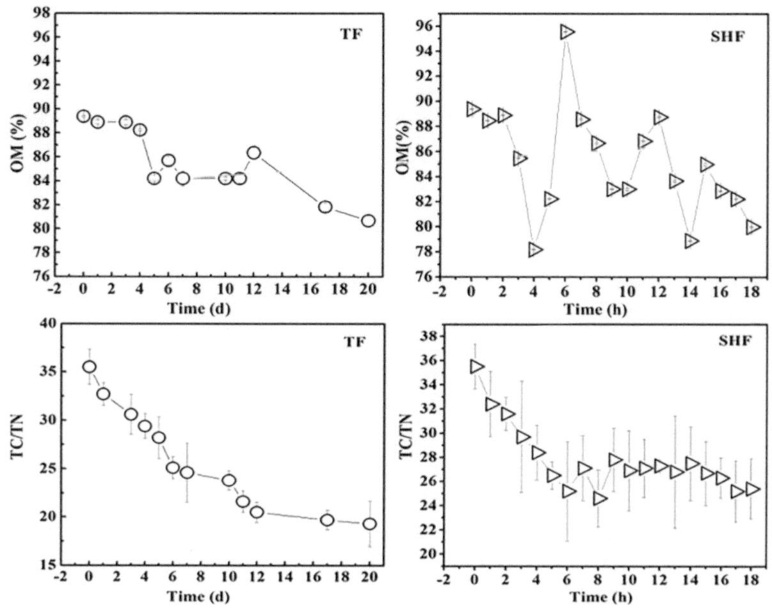

Fig. 3.2 (continued)

of reaction, the carbon mineralization rate and the nitrogen fixation rate reached a dynamic equilibrium (Masakazu and Aoyama 1993).

The changes in pH, EC, and TC/TN of SHF were similar to those reported in existing literature (Jiang et al. 2015), which reported a conditioner product fermented within 96 h without inoculating an exogenous composite microbial agent. This indicates that an external heat source and mechanical stirring can accelerate the humification process, while inoculating exogenous microorganisms can further shorten the fermentation cycle by degrading recalcitrant large-molecule proteins and fatty substances into small-molecule substances.

3.2.2 Morphological Characterization

3.2.2.1 Particle Size Composition

Particle grading experiments can be used to evaluate the particle size distribution of granular materials (Gabhane et al. 2012b). The particle size distribution of TF and SHF is shown in Fig. 3.3. The particle size of TF was concentrated between 2 and 4 mm, accounting for 62.14% of the total material volume, while particles < 1 mm accounted for 7.44%. The particle size of SHF was concentrated below 3 mm, with the mass of particulate material with a particle size < 1 mm accounting for 69.37% of the total material. Generally, 90% of the material in a fertilizer or soil conditioner applied to the soil should be able to pass through a 12.6 mm sieve (Gabhane et al. 2012b). Both TF and SHF meet this standard requirement. Compared to traditional fermentation, high-temperature hydrothermal hydrolysis fermentation has a better degradation effect on food waste. This indicates that the inoculation of a composite microbial agent, the alternating forward and reverse stirring method, and the external heat source accelerated the physical collision and biodegradation of the material. The fine particulate material < 1 mm has a large specific surface area and is rich in small molecular available carbon. This not only facilitates the attachment and reproduction of microorganisms but also has a "priming effect" on the indigenous microbial communities of barren soil when applied (Hou et al. 2017).

As shown in Table 3.2, the total nitrogen, total phosphorus, and total potassium contents of both TF and SHF increased with increasing particle size, while pH and electrical conductivity decreased with increasing particle size. In TF, the total carbon, total nitrogen, total phosphorus, total potassium, pH, and electrical conductivity of materials with a particle size greater than 5 mm and between 1–3 mm were significantly higher than those of materials with the same particle size in the SHF soil conditioner. In SHF, the material with a particle size of less than 1 mm had higher total carbon, total nitrogen, total phosphorus, and total potassium contents relative to the material of the same particle size in TF. When applied to the soil, this is beneficial for increasing soil nutrient content.

During the fermentation process, a TC/TN < 20 indicates that the fermentation product has reached maturity (Tang et al. 2003). In TF, the TC/TN of material with a particle size < 2 mm was < 20, and in SHF, the TC/TN of material with a particle size < 4 mm was < 20. This is because material with a particle size > 5 mm often

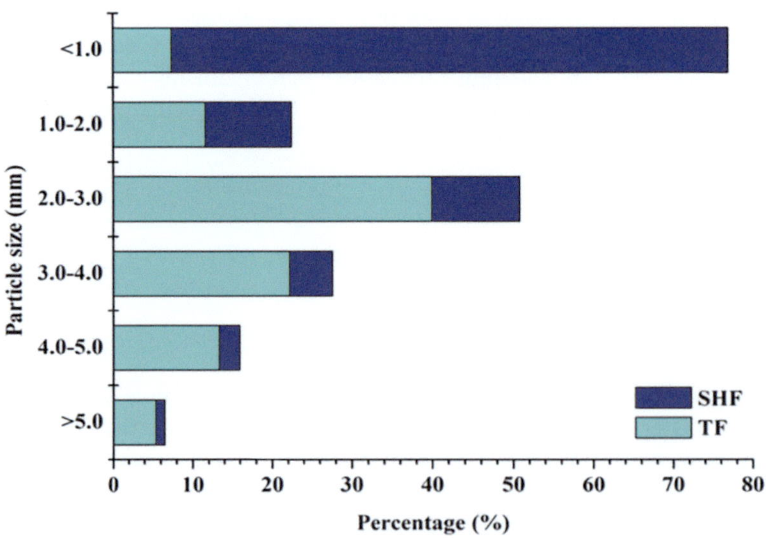

Fig. 3.3 Particle size distribution of TF and SHF

Table 3.2 Physicochemical properties of different particle sizes of TF and SHF

Particle size (mm)	pH		EC (mS cm^{-1})		TC (g kg^{-1})		TN (g kg^{-1})	
	TF	SHF	TF	SHF	TF	SHF	TF	SHF
> 5.0	6.32 ± 0.02	6.11 ± 0.03	2.88 ± 0.05	2.08 ± 0.04	432.55 ± 6.65	384.57 ± 5.69	18.60 ± 0.37	17.66 ± 0.35
4.0–5.0	6.33 ± 0.03	6.94 ± 0.04	2.84 ± 0.08	2.32 ± 0.07	416.38 ± 6.33	353.62 ± 5.07	20.32 ± 0.41	19.55 ± 0.39
3.0–4.0	6.85 ± 0.03	7.33 ± 0.03	2.96 ± 0.05	2.77 ± 0.06	387.43 ± 5.75	321.68 ± 4.43	16.35 ± 0.33	18.42 ± 0.37
2.0–3.0	7.16 ± 0.02	7.69 ± 0.02	3.39 ± 0.07	2.99 ± 0.08	385.66 ± 5.71	330.59 ± 4.61	16.55 ± 0.33	17.37 ± 0.35
1.0–2.0	8.25 ± 0.06	8.11 ± 0.05	3.57 ± 0.09	3.02 ± 0.05	331.57 ± 4.63	327.84 ± 4.56	16.97 ± 0.34	17.49 ± 0.35
< 1.0	8.38 ± 0.03	8.55 ± 0.02	4.16 ± 0.06	3.58 ± 0.04	306.84 ± 4.13	310.44 ± 4.21	15.61 ± 0.31	15.86 ± 0.32

Particle size (mm)	TC/TN		TP (g kg^{-1})		TK (g kg^{-1})	
	TF	SHF	TF	SHF	TF	SHF
> 5.0	23.26 ± 0.36	21.78 ± 0.32	0.46 ± 0.01	0.55 ± 0.02	10.37 ± 0.21	9.62 ± 0.19
4.0–5.0	20.49 ± 0.31	18.09 ± 0.26	0.41 ± 0.02	0.48 ± 0.03	10.11 ± 0.20	9.51 ± 0.19
3.0–4.0	23.70 ± 0.35	17.46 ± 0.24	0.39 ± 0.03	0.40 ± 0.02	10.09 ± 0.21	9.06 ± 0.18
2.0–3.0	23.30 ± 0.35	19.03 ± 0.27	0.40 ± 0.02	0.35 ± 0.01	10.1 ± 0.20	9.02 ± 0.18
1.0–2.0	19.54 ± 0.27	18.74 ± 0.26	0.38 ± 0.02	0.38 ± 0.03	9.55 ± 0.19	8.33 ± 0.17
< 1.0	19.66 ± 0.26	19.57 ± 0.27	0.33 ± 0.03	0.35 ± 0.02	9.84 ± 0.20	8.31 ± 0.17

experiences localized insufficient oxygen supply, leading to incomplete maturation of the material (Jiang et al. 2015).

3.2.2.2 Molecular Weight Fractionation of Dissolved Organic Matter (DOM)

DOM has a relatively complex chemical structure and amorphous molecular weight, and changes in its structure and composition can reflect the fermentation process and the degree of material humification (Said-Pullicino et al. 2007b; Wang et al. 2013; Wei et al. 2016). As shown in Fig. 3.4, compared to the middle stage of fermentation, the proportion of DOM with a molecular weight (MW) < 1 kDa decreased in both TF and SHF at the end of fermentation, while the proportion of material with an MW > 5 kDa increased. This indicates that during the fermentation process, small-molecule organic matter polymerized into large-molecule components, and the fermentation proceeded in a direction that promoted the humification process.

In the original material, the component of DOM with a molecular weight between 1 and 5 kDa accounted for 11.5%. After fermentation, the proportion of this component in TF decreased to 5.22%, while in SHF it increased to 12.13%. The component with an MW > 5 kDa was the main material in both TF and SHF, accounting for 82.57% and 63.48% in the TF and SHF soil conditioners, respectively. During the fermentation process, DOM with a molecular weight greater than 5 kDa is mostly humic carbon-like inert organic components and is considered a standard for the maturity of the fermentation product; that is, the higher the content of the MW >

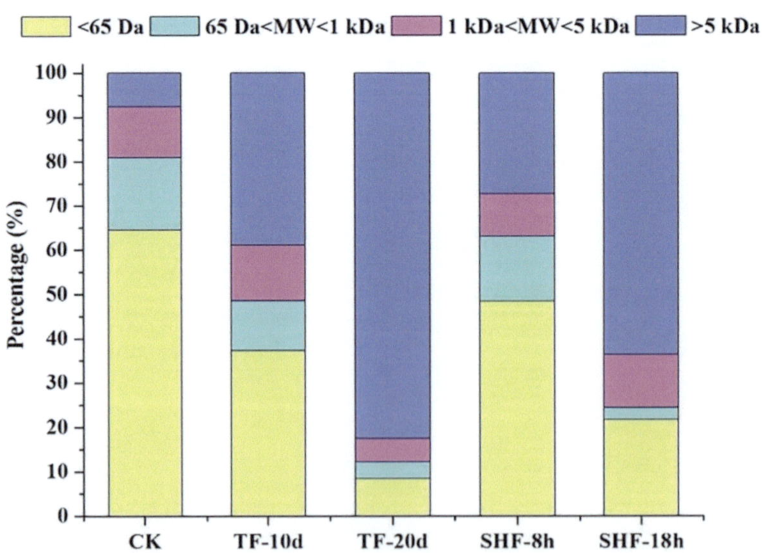

Fig. 3.4 Molecular weight distribution of DOM components in TF and SHF

5 kDa component in the fermented material, the higher the degree of humification (Wei et al. 2014). The MW > 5 kDa component can provide slow-release nutrients to promote the formation of aggregates (Wei et al. 2016, 2014), while the MW < 1 kDa component is usually small molecular available carbon that can provide a direct nutritional source for the growth of microorganisms and crops (Hou et al. 2017; Weiss et al. 1991). At the end of fermentation, the humic carbon and readily available carbon components in SHF accounted for 63.48% and 24.39%, respectively, representing the main carbon forms in the SHF conditioner. This carbon distribution allows that, when SHF is applied to barren soil, its macromolecular humic carbon can serve as a substrate for the condensation of barren soil aggregates and the formation of humus, while the small molecular available carbon can provide energy and a nutritional source for increasing the abundance and diversity of the microbial community.

3.2.2.3 Structural Characteristics

Scanning results showed that the morphological structure of TF and SHF changed significantly during the fermentation process. As shown in Fig. 3.5, compared with the initial material (CK) and the material in the middle stage of fermentation (10 d), the morphological structure of TF changed significantly at the end of fermentation (20 d), and the porosity of the material increased markedly. Compared with TF during the fermentation process, the internal fibrous structure of SHF was more degraded, the specific surface area of the material was larger, and more microorganisms were attached. This is attributed to the bioaugmented degradation effect of exogenous microorganisms on the fermentation material which, under the influence of the high-temperature dynamic stirring environment, caused a more thorough degradation of SHF.

3.2.3 Three-Dimensional Fluorescence Spectroscopy Analysis

The fermentation process is a process of humus synthesis. The humification rate and functional group structure of humic substances are often used to evaluate the stability and degree of maturity of the fermentation product. Three-dimensional fluorescence (EEM) technology was used to study the structural characteristics of DOM and humic substances during the fermentation process to reflect the composting process and maturity (Lv et al. 2014; Zhang et al. 2016a; Fu et al. 2005). As shown in Fig. 3.6a, the three-dimensional fluorescence components of DOM during the fermentation process were divided into 5 regions. Regions I, II, III, IV, and V represent tyrosine-like substances, tryptophan-like substances, fulvic acid-like substances, soluble microbial by-products, and humic acid-like substances, respectively. The Pi,n value corresponding to each region can reflect the relative content of its corresponding

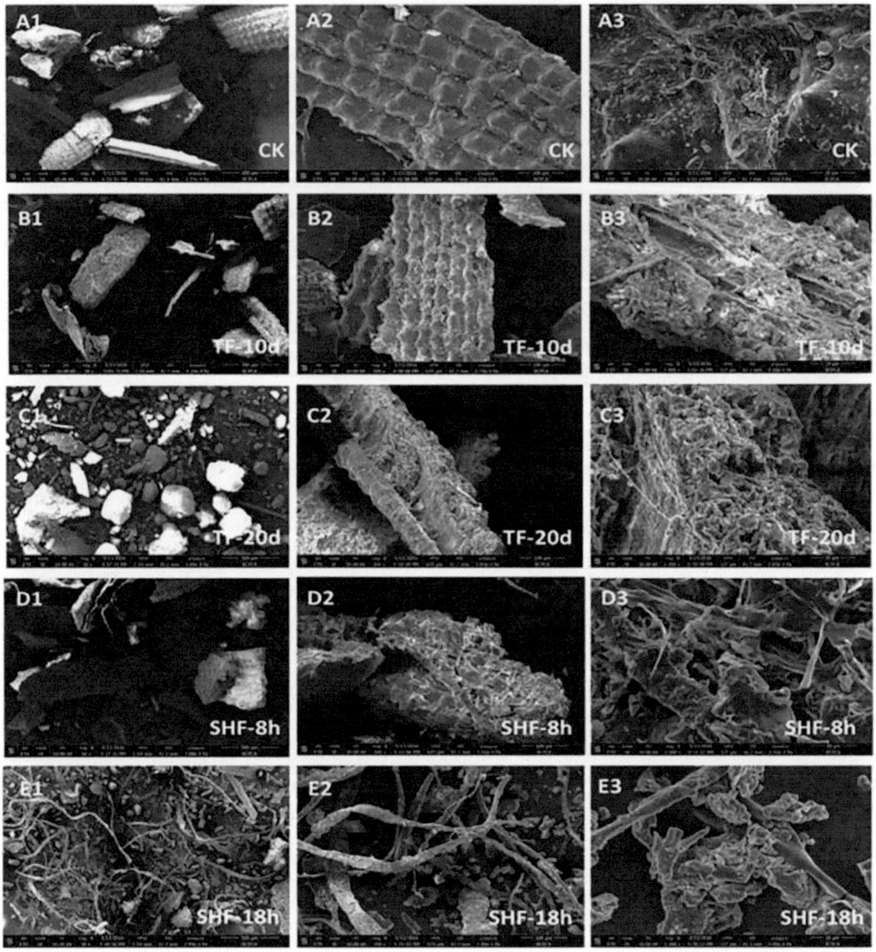

Fig. 3.5 Morphological structure of TF and SHF. **a** CK, original material composed of food waste and rice husk; **b** TF-10 d, traditional fermentation mid-stage (10 d); **c** TF-20 d, traditional fermentation end-stage (20 d); **d** SHF-8 h, short-term hydrothermal hydrolysis fermentation mid-stage (8 h); E: SHF-18 h, short-term hydrothermal hydrolysis fermentation end-stage (18 h); Left column: 50 × magnification; Middle column: 200 × magnification; Right column: 1000 × magnification

substance (Chen et al. 2003; Wang et al. 2013) (Fig. 3.6b). The excitation/emission wavelengths corresponding to the maximum fluorescence peak positions in Regions I and II were 224/308 nm and 278/330 nm, respectively. According to previous research reports, these two fluorescence peaks are mainly protein-like fluorescence peaks, representing the tyrosine-like fluorescence peak (Peak I) and the tryptophan-like fluorescence peak (Peak II), respectively (Ahmad and Reynolds 1999; Lv et al. 2013). Both of the two types of substances are related to the metabolic activities of microorganisms (Chen et al. 2003). The excitation/emission wavelength corresponding to the maximum fluorescence peak position of Peak V was 348/430 nm. He et al. (2016) believe that the fluorescence peak in the region near the excitation/emission wavelength of 288/455 nm represents humic-like substances, and the change in its peak intensity is directly proportional to the change in the content of humic-like substances. Changes in these peak intensities can indicate the humification process of the compost. The presence of the Region V fluorescence peak and the signal change of its peak intensity can indicate the stability of the material during the fermentation process. From Fig. 3.6a, it can be seen that during the fermentation of TF and SHF, the fluorescence peak signal intensity in both Region I and Region II weakened, indicating that protein-like substances were degraded and utilized (Zhang et al. 2016b). As fermentation proceeded, the fluorescence peak intensity of Region V gradually increased, humic-like substances were generated, and the stability of the material increased (Marhuenda-Egea et al. 2007).

As shown in Fig. 3.6b, the volume integral ratios of the three-dimensional fluorescence Regions I, II, and IV for TF and SHF gradually decreased as fermentation proceeded. In CK, the volume integral ratios of the above three fluorescence regions were 19.2%, 14.8%, and 35.3%, respectively. After fermentation, the volume integrals of these three regions in TF decreased to 7.9%, 9.8%, and 26%, respectively. In SHF, they decreased to 16.1%, 11.2%, and 29.7%, respectively. The volume integral ratio of Region IV in SHF was 7% higher than in TF. The substances in this region are mostly microbial metabolic by-products. As small-molecule carbon, they can provide a direct nutritional source for crop growth. In addition, microbial metabolic by-products are necessary cementing agents for the stabilization of soil aggregates and can promote an increase in soil organic matter content. The volume integral ratio of Region V in CK was 20.0%. After fermentation, it increased to 54.4% and 33.5% in TF and SHF, respectively. This result shows that as fermentation proceeded, humic acid-like substances gradually became the main component of DOM. This type of substance is usually formed from organo-mineral complexes, microaggregates, and macroaggregates, has a high degree of molecular condensation and humification, and a complex molecular structure, is not easily degraded, and can play the role of a slow-release nutrient in long-term application, which is beneficial for long-term soil water and fertilizer retention and carbon sequestration (He et al. 2013).

3.2.4 Bacterial Community Structure and Composition

The composition of the microbial community in a conditioner is a primary factor influencing its effectiveness. Using high-throughput sequencing as a technical tool, the dominant microbial community structures of TF and SHF were analyzed, with the results shown in Fig. 3.7. In the control (CK) treatment, *Firmicutes*, *Proteobacteria*, and *Acidobacteria* were the dominant phyla. After fermentation, *Bacteroidetes* replaced *Acidobacteria* as a dominant microorganism in both TF and SHF.

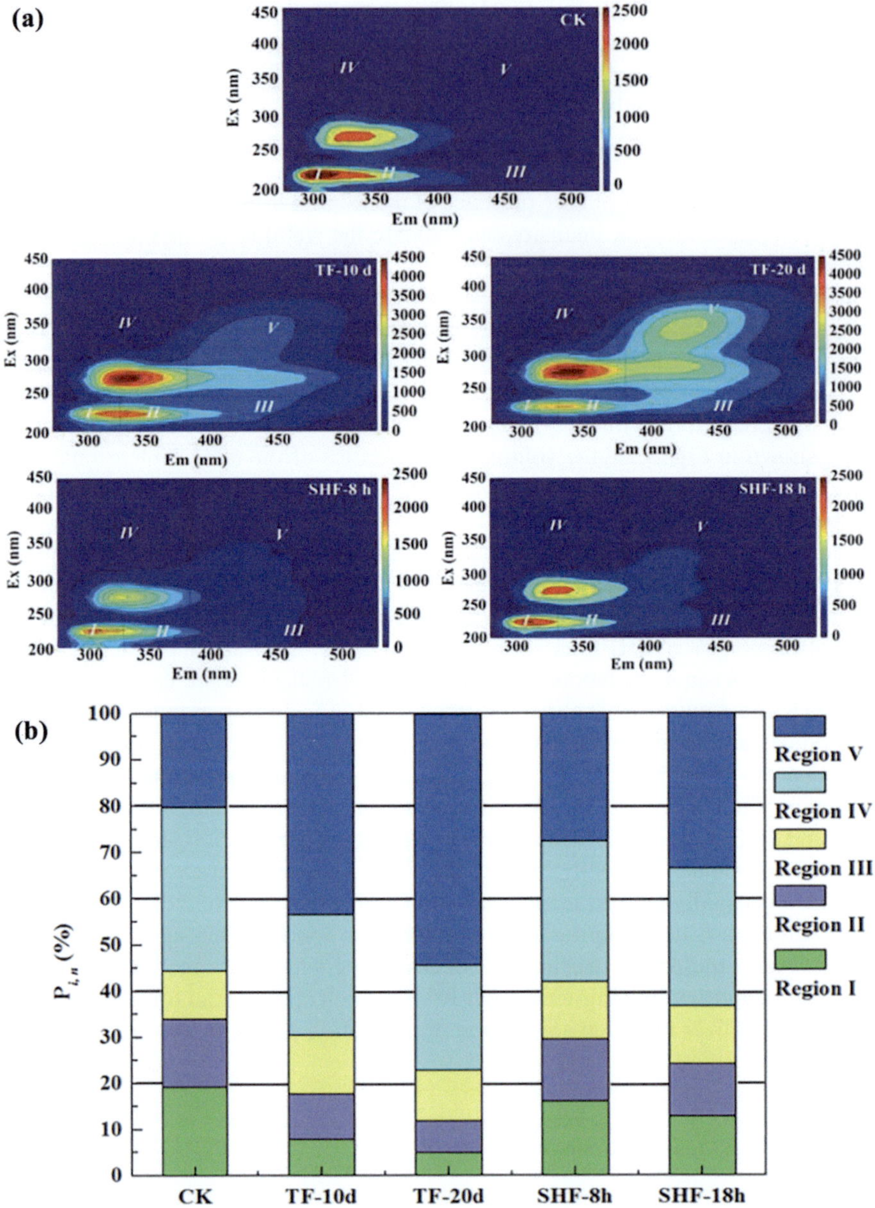

Fig. 3.6 **a** Changes in the three-dimensional fluorescence spectra of TF and SHF; **b** The volume integral ratio of three-dimensional fluorescence regions at each stage of TF and SHF fermentation

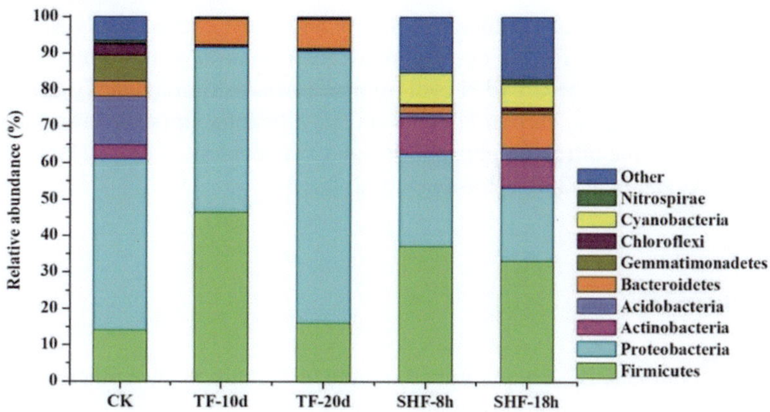

Fig. 3.7 Distribution of major microbial phyla in TF and SHF

The relative abundance of *Proteobacteria* in the CK was 46.93%. After fermentation, its relative abundance increased to 74.66% in TF and decreased to 20.22% in SHF. The relative abundance of *Nitrospirae* changed from 0.95% in the CK treatment group to 0.02% and 1.32% in TF and SHF, respectively. Certain genera within *Nitrospirae* can promote nitrogen transformation and cycling through microbial metabolism (Wang et al. 2012; Yan et al. 2015).

Actinobacteria are widely distributed in terrestrial environments (Stach et al. 2003). After fermentation, the relative abundance of *Actinobacteria*, which was 3.77% in the CK treatment, decreased to 0.45% in TF and increased to 7.79% in SHF. Actinobacteria are involved in the transformation of soil organic matter and play a significant role in degrading recalcitrant components (Schrempf 2001). Furthermore, Actinobacteria can also influence the nitrogen fixation process in non-leguminous plants (Benson and Silvester 1993). The relative abundances of *Acidobacteria*, *Gemmatimonadetes*, and *Chloroflexi* decreased in both TF and SHF, while the relative abundance of *Bacteroidetes* increased significantly. After fermentation, the relative abundance of *Bacteroidetes* in SHF was 1.2 times that in TF. This is likely related to the composite microbial inoculant, which is rich in bacilli, that was introduced during the SHF fermentation process. However, the difference between the two was not significant. It is speculated that the exogenous microorganisms played a primary role in the initial stage of SHF fermentation. At the 8-h mark of fermentation, their relative abundance decreased to 1.33% when the temperature was 71.61 °C, indicating that most *Bacteroidetes* are not tolerant to high-temperature environments. However, after the fermentation concluded and the material temperature dropped to 41.63 °C, their relative abundance increased to 9.27%. This suggests that the phylum *Bacteroidetes* readapted to the changes in the external environment, and its growth and reproduction were positively stimulated.

The relative abundance of *Cyanobacteria* in the CK was 0.22%. On days 10 and 20 of the traditional fermentation, the relative abundance of this bacterium in TF decreased to 0.01% and 0.02%. At 8 h and 18 h of the hydrothermal hydrolysis fermentation, its relative abundance was 8.56% and 6.29%, respectively. *Cyanobacteria* are very important for the cycling and transformation of nutrients, especially substances involved in the C and N cycling processes. Additionally, *Cyanobacteria* can increase soil organic carbon content and significantly improve crop yield and quality, as well as the nitrogen, phosphorus, and zinc content in crops (Jaiswal et al. 2009; Karthikeyan et al. 2009; Prasanna et al. 2013).

3.3 Summary

This chapter coupled high-temperature dynamic fermentation, hydrothermal hydrolysis, and a bio-augmentation inoculation process to prepare a synergistic and high-efficiency conditioner. Using organic fertilizer produced via a traditional fermentation method as a reference, this study characterized the conditioner's physicochemical, morphological, and spectroscopic features, as well as its microbial community composition. The main conclusions are as follows:

(1) During the fermentation process, the various physicochemical parameters (e.g., electrical conductivity, organic matter, TC/TN, moisture content) of the synergistic and high-efficiency conditioner were similar to those of the traditional fertilizer. The peak temperature in traditional composting occurred on day 7 at 62 °C, whereas the synergistic and high-efficiency conditioner reached its peak temperature of 71.94 °C at the 14th hour of fermentation. The process combining high-temperature dynamic fermentation, hydrothermal hydrolysis, and bio-augmentation inoculation promoted the material's humification process and shortened the fermentation period. After fermentation, the synergistic and high-efficiency conditioner had a pH of 7.1, a moisture content of 22%, and an organic matter content of 82%. Within 18 h, it met the application standards for a soil conditioner.

(2) The proportion of particles < 1 mm in the synergistic and high-efficiency conditioner was 44%, which was 5.9 times that of the traditional fertilizer. Its larger specific surface area is conducive to microbial attachment and proliferation. The proportion of labile carbon (MW < 1 kDa) in the synergistic and high-efficiency conditioner was 24.39%, twice that of the traditional fertilizer. This small-molecule labile carbon can provide energy and nutrient sources to enhance the abundance and diversity of the microbial community. The proportion of humic carbon (MW > 5 kDa) was 63.48%, which can serve as a substrate for the polycondensation of aggregates and the formation of humus in impoverished soils. Spectroscopic characterization showed that microbial metabolic byproducts in the synergistic and high-efficiency conditioner were 7% higher than in the traditional fertilizer, and its carbohydrate-derived carbon was twice as high. This indicates that the synergistic and high-efficiency conditioner has greater potential to contribute to the binding force of soil aggregates and the turnover of the soil organic carbon pool.

(3) *Firmicutes*, *Proteobacteria*, and *Bacteroidetes* were the dominant microorganisms in both the synergistic and high-efficiency conditioner and the traditional fertilizer. Bio-augmentation and high-temperature hydrothermal hydrolysis fermentation had a positive stimulatory effect on the growth and reproduction of *Nitrospirae*, *Actinobacteria*, and *Cyanobacteria*. This suggests that the synergistic and high-efficiency conditioner has the potential to promote soil nitrogen transformation, organic carbon sequestration, and crop growth.

References

Ahmad SR, Reynolds DM (1999) Monitoring of water quality using fluorescence technique: prospect of on-line process control. Water Res 33(9):2069–2074

Benson DR, Silvester WB (1993) Biology of Frankia strains, actinomycete symbionts of actinorhizal plants. Microbiol Rev 57(2):293

Chen W, Westerhoff P, Leenheer JA, Booksh K (2003) Fluorescence excitation-emission matrix regional integration to quantify spectra for dissolved organic matter. Environ Sci Technol 37(24):5701

Chen Y, Yu F, Liang S, Wang Z, Liu Z, Xiong Y (2014) Utilization of solar energy in sewage sludge compositing: fertilizer effect and application. Waste Manage 34(11):2014–2021

Chiew YL, Spangberg J, Baky A, Hansson P-A, Jonsson H (2015) Environmental impact of recycling digested food waste as a fertilizer in agriculture—a case study. Resour Conserv Recycl 95:1–14

Coelho C, Cavani L, ter Halle A, Guyot G, Ciavatta C, Richard C (2011) Rates of production of hydroxyl radicals and singlet oxygen from irradiated compost. Chemosphere 85(4):630–636

Fu P, Liu C, Wu F (2005) Three-dimensional excitation emission matrix fluorescence spectroscopic characterization of dissolved organic matter. Spectrosc Spectr Anal 25(12):2024–2028

Gabhane J, William SPMP, Bidyadhar R, Bhilawe P, Anand D, Vaidya AN, Wate SR (2012) Additives aided composting of green waste: effects on organic matter degradation, compost maturity, and quality of the finished compost. Biores Technol 114:382–388

He L, Huang GH, Lu H (2011) Greenhouse gas emissions control in integrated municipal solid waste management through mixed integer bilevel decision-making. J Hazard Mater 193:112–119

He X, Xi B, Gao R, Zhang H, Dang Q, Li D, Huang C (2016) Insight into the composition and degradation potential of dissolved organic matter with different hydrophobicity in landfill leachates. Chemosphere 144:75–80

He X, Xi B, Li X, Pan H, An D, Bai S, Li D, Cui D (2013) Fluorescence excitation-emission matrix spectra coupled with parallel factor and regional integration analysis to characterize organic matter humification. Chemosphere 93(9):2208–2215

Hou J, Li M, Xi B, Tan W, Jie D, Yan H, Liu D, Liu H (2017) Short-duration hydrothermal fermentation of food waste: preparation of soil conditioner for amending organic-matter-impoverished arable soils. Environ Sci Pollut Res 9:1–15

Jaiswal P, Prasanna R, Nayak S, Sood A, Suseela MR (2009) Characterization of rhizo-cyanobacteria and their associations with wheat seedlings. Egypt J Biol 10(1):20–27

Jiang Y, Ju M, Li W, Ren Q, Liu L, Chen Y, Yang Q, Hou Q, Liu Y (2015) Rapid production of organic fertilizer by dynamic high-temperature aerobic fermentation (DHAF) of food waste. Biores Technol 197:7–14

Karthikeyan N, Prasanna R, Sood A, Jaiswal P, Nayak S, Kaushik B (2009) Physiological characterization and electron microscopic investigation of cyanobacteria associated with wheat rhizosphere. Folia Microbiol 54(1):43–51

Kuo W, Cheng K (2007) Use of respirometer in evaluation of process and toxicity of thermophilic anaerobic digestion for treating kitchen waste. Biores Technol 98(9):1805–1811

Lashermes G, Barriuso E, Houot S (2012) Dissipation pathways of organic pollutants during the composting of organic wastes. Chemosphere 87(2):137–143

Li W, Chen S, Xu Z, Li Y, Shuang C, Li A (2014) Characterization of dissolved organic matter in municipal wastewater using fluorescence PARAFAC analysis and chromatography multi-excitation/emission scan: a comparative study. Environ Sci Technol 48(5):2603–2609

Li Z, Lu H, Ren L, He L (2013) Experimental and modeling approaches for food waste composting: a review. Chemosphere 93(7):1247–1257

Lv B, Xing M, Yang J, Qi W, Lu Y (2013) Chemical and spectroscopic characterization of water extractable organic matter during vermicomposting of cattle dung. Biores Technol 132(2):320–326

Lv B, Xing M, Zhao C, Yang J, Xiang L (2014) Towards understanding the stabilization process in vermicomposting using PARAFAC analysis of fluorescence spectra. Chemosphere 117(1):216

Marhuenda-Egea F, Martínez-Sabater E, Jordá J, Moral R, Bustamante MA, Paredes C, Pérez-Murcia M (2007) Dissolved organic matter fractions formed during composting of winery and distillery residues: evaluation of the process by fluorescence excitation–emission matrix. Chemosphere 68(2):301

Masakazu A, Tomohiro N (1993) Microbial biomass nitrogen and mineralization-immobilization processes of nitrogen in soils incubated with various organic materials. Soil Sci Plant Nutr 39(1):23–32

Prasanna R, Chaudhary V, Gupta V, Babu S, Kumar A, Singh R, Shivay Y, Nain L (2013) Cyanobacteria mediated plant growth promotion and bioprotection against Fusarium wilt in tomato. Eur J Plant Pathol 136(2):337–353

Said-Pullicino D, Erriquens F, Gigliotti G (2007a) Changes in the chemical characteristics of water-extractable organic matter during composting and their influence on compost stability and maturity. Biores Technol 98(9):1822–1831

Schrempf H (2001) Recognition and degradation of chitin by streptomycetes. Antonie Van Leeuwenhoek 79(3–4):285–289

Stach J, Maldonado L, Ward A, Goodfellow M, Bull A (2003) New primers for the class Actinobacteria: application to marine and terrestrial environments. Environ Microbiol 5(10):828–841

Sundberg C, Smars S, Jonsson H (2004) Low pH as an inhibiting factor in the transition from mesophilic to thermophilic phase in composting. Biores Technol 95(2):145–150

Tang J, Inoue Y, Yasuta T, Yoshida S, Katayama A (2003) Chemical and microbial properties of various compost products. Soil Science Plant Nutrition 49(2):273–280

Wang K, Li W, Gong XJ, Li Y, Wu C, Ren N (2013) Spectral study of dissolved organic matter in biosolid during the composting process using inorganic bulking agent: UV–vis, GPC, FTIR and EEM. Int Biodeterior Biodegradation 85(7):617–623

Wang X, Selvam A, Wong J (2016) Influence of lime on struvite formation and nitrogen conservation during food waste composting. Biores Technol 217:227–232

Wang Y, Sheng H, He Y, Wu J, Jiang Y, Tam N, Zhou H (2012) Comparison of the levels of bacterial diversity in freshwater, intertidal wetland, and marine sediments by using millions of illumina tags. Appl Environ Microbiol 78(23):8264

Wei Z, Wang X, Zhao X, Xi B, Wei Y, Zhang X, Zhao Y (2016) Fluorescence characteristics of molecular weight fractions of dissolved organic matter derived from composts. Int Biodeterior Biodegrad 113:187–194

Wei Z, Zhang X, Wei Y, Wen X, Shi J, Wu J, Zhao Y, Xi B (2014) Fractions and biodegradability of dissolved organic matter derived from different composts. Biores Technol 161(6):179–185

Weiss MS, Abele U, Weckesser J, Welte W, Schiltz E, Schulz G (1991) Molecular architecture and electrostatic properties of a bacterial porin. Science 254(5038):1627–1630

Yan N, Marschner P, Cao W, Zuo C, Qin W (2015) Influence of salinity and water content on soil microorganisms. Int Soil Water Conserv Res 3(4):316–323

Zaher U, Higgins S, Carpenter-Boggs L (2016) Interactive life cycle assessment framework to evaluate agricultural impacts and benchmark emission reduction credits from organic management. J Clean Prod 115:182–190

Zhang S, Chen Z, Wen Q, Ma J, He Z (2016a) Assessment of maturity during co-composting of penicillin mycelial dreg via fluorescence excitation-emission matrix spectra: characteristics of chemical and fluorescent parameters of water-extractable organic matter. Chemosphere 155:358–366

Zhang S, Chen Z, Wen Q, Zheng J (2016b) Assessing the stability in composting of penicillin mycelial dreg via parallel factor (PARAFAC) analysis of fluorescence excitation–emission matrix (EEM). Chem Eng J 299:167–176

Chapter 4
Role of the Synergistic and High-Efficiency Soil Conditioner in Promoting the Humification of Impoverished Soil

Abstract Soil classification systems define soil with an organic matter content of less than 1% as having extremely low fertility, i.e., as impoverished soil. In China, the area of impoverished soil accounts for 2/5 of the total arable land. One of the key priorities for improving the quality of impoverished arable land is to increase soil organic matter content and promote the formation of humic aggregate structures. Soil aggregates are the basic structural units for the formation of soil humus and the formation rate of soil humus is a critical indicator that directly determines the level of soil organic matter. Investigating the changing characteristics of the precursor substances for soil humus formation is significant for studying the formation of soil humus and changes in soil fertility. This chapter, using traditional organic fertilizer (TF) and chemical fertilizer (HF) as reference objects, clarifies the mechanism by which the synergistic and high-efficiency conditioner (SHF) affects the formation of humus in infertile arable soil and evaluates its effect on the management of the carbon pool in low-quality arable land.

4.1 Effect of the Synergistic Conditioner on Humus Formation in Impoverished Soil

4.1.1 Changes in Soil Physicochemical Properties

Their particle size distribution and nutrient content are decisive factors in the stability of soil humus. As the main component of soil organic matter, the formation rate of soil humus is a critical indicator that directly determines the level of soil organic matter and is a primary marker of soil fertility (Badía-Villas and Girona-García 2017; Kostenko, 2017). There are several hypotheses regarding the formation of soil humus, with the sugar-amine condensation theory and the polyphenol theory being the two most widely accepted in the academic community. The sugar-amine condensation theory posits that polysaccharides, reducing sugars, and amino acids produced by microbial metabolism are the precursor substances for soil humus formation; these compounds form humic polymers through non-enzymatic polymerization. The

J. Hou et al., *Food Waste Valorization: Preparation, Properties, and Application of the Synergistic Soil Conditioners*, https://doi.org/10.1007/978-981-95-5362-4_4

polyphenol theory suggests that after crop residues are decomposed by microorganisms, they form phenolic, hydroxyphenolic, and amino acid-type compounds, which can then form humic substances through chemical oxidation and polymerization (Dou Sen 2010). Regarding the humification process, the aforementioned pathways may occur interactively and influence each other; their relative importance simply varies depending on environmental conditions. Therefore, investigating the changing characteristics of the precursor substances for soil humus formation is significant for studying the formation of soil humus and changes in soil fertility. This chapter, using traditional organic fertilizer (TF) and chemical fertilizer (HF) as reference objects, clarifies the mechanism by which the synergistic and high-efficiency conditioner (SHF) affects the formation of humus in infertile arable soil and evaluates its effect on the management of the carbon pool in low-quality arable land.

The experiment was conducted at the Groundwater and Environmental Systems Engineering Innovation Base of the Chinese Academy of Environmental Sciences, located in the northern part of the North China Plain (N 40° 11′ 24″, E 116° 35′ 14″). The area has a temperate continental semi-humid monsoon climate, with an average annual temperature of 11.5 °C, 2746 annual sunshine hours, an annual relative humidity of 58%, a 195-day frost-free period, and an average annual rainfall of 610 mm. The soil type at the experimental site is fluvo-aquic soil, with the cultivated layer composed of 10.80% clay, 74.11% silt, and 15.09% sand. Before the experiment began, the land was not cultivated, irrigated, or treated with any fertilizer, although it was regularly leveled and mechanically weeded. The preparation methods for the tested synergistic and high-efficiency conditioner (SHF) and organic fertilizer (TF) are described in Chap. 3 of this book. The tested chemical fertilizer (HF) was a commercial NPK ternary compound fertilizer, with total nitrogen, phosphorus pentoxide, and potassium oxide each accounting for 15%. The basic properties of the soil are shown in Table 4.1.

As shown in Fig. 4.1, compared to the initial background pH of the soil, the pH of the soil without the addition of exogenous nutrients did not change significantly. This indicates that in an environment without cultivation, fertilization, or crop growth, the soil's acidity and alkalinity did not change over a short period. After the application of TF and SHF, the soil pH decreased slightly, which is related to the fact that TF itself has a pH < 7. Additionally, as the experiment progressed, microorganisms and various biological enzymes in the impoverished soil released a

Table 4.1 Basic properties of the test soil A

Physicochemical property (unit)	Value
pH	8.58 ± 0.74
Organic matter (g kg^{-1})	5.28 ± 0.06
Total nitrogen (g kg^{-1})	0.50 ± 0.02
Alkali-hydrolyzable nitrogen (mg kg^{-1})	36.08 ± 3.28
Available phosphorus (mg kg^{-1})	12.18 ± 1.11
Available potassium (mg kg^{-1})	50.11 ± 2.66

large number of microbial metabolic byproducts and small-molecule acids through a series of mineralization and decomposition processes, which is another reason for the slight decrease in soil pH. After the application of HF, the soil pH increased slightly, which is related to the properties of the chemical fertilizer itself. The application of exogenous nutrients (TF, SHF, HF) increased soil nutrient content to varying degrees. Among them, the chemical fertilizer had the most significant effect on increasing nutrient content, especially the available potassium content. This shows that within the experimental period, the supply intensity of available nitrogen, phosphorus, and potassium from the chemical fertilizer was higher than that from the traditional fertilizer and the rapid hydrothermal hydrolysis fermentation conditioner. The application of TF, SHF, and HF increased soil organic matter by 0.18 g kg^{-1}, 0.62 g kg^{-1}, and 0.14 g kg^{-1}, respectively. This demonstrates that the synergistic and high-efficiency conditioner increases the reserves of the soil organic carbon pool in impoverished soil significantly more than traditional fertilization methods.

4.1.2 Characteristics of Aggregate Size and Nutrient Distribution

4.1.2.1 Distribution of Soil Aggregate Sizes

The distribution of different size fractions of soil aggregates is greatly influenced by the application of exogenous nutrients (TF, SHF, HF). As shown in Table 4.2, the proportions of the different aggregate size fractions, from largest to smallest, were: > 2 mm aggregates, 0.5–1 mm aggregates, 1–2 mm aggregates, < 0.05 mm aggregates, 0.25–0.5 mm aggregates, and 0.05–0.25 mm aggregates.

Compared to the initial soil (CK0), the content of < 0.05 mm size fraction aggregates in CK-S increased significantly, while the content of 0.5–2 mm and 0.05–0.25 mm size fraction aggregates decreased slightly. The particle size distribution of soil aggregates is an important indicator for evaluating soil erodibility. Research has shown that the higher the proportion of fine silt particles in the soil, the greater its wind erodibility (Chepil 1951). Compared to CK-S, the content of > 2 mm and < 0.05 mm size fraction aggregates in TF-S were both significantly lower, whereas the percentage of 0.05–2 mm size fraction aggregates increased markedly, which is consistent with the changes in aggregate size fractions in SHF-S. The research of Kristiansen et al. (2006) pointed out that the 0.25–1 mm fraction is the most abundant aggregate proportion in arable soil (28.5–62.5%), which is similar to the conclusions of this study. Some research has indicated that a greater number of > 1 mm size fraction aggregates lead to a lower capacity for nutrient element retention and a greater potential for soil nutrient element release (Wang et al. 2017). In SHF-S, the content of 0.05–1 mm size fraction aggregates increased by 18.35% and 10.25% relative to CK and TF-S, respectively, indicating that the application of SHF significantly increased the content of 0.05–1 mm size fraction aggregates in the soil, which

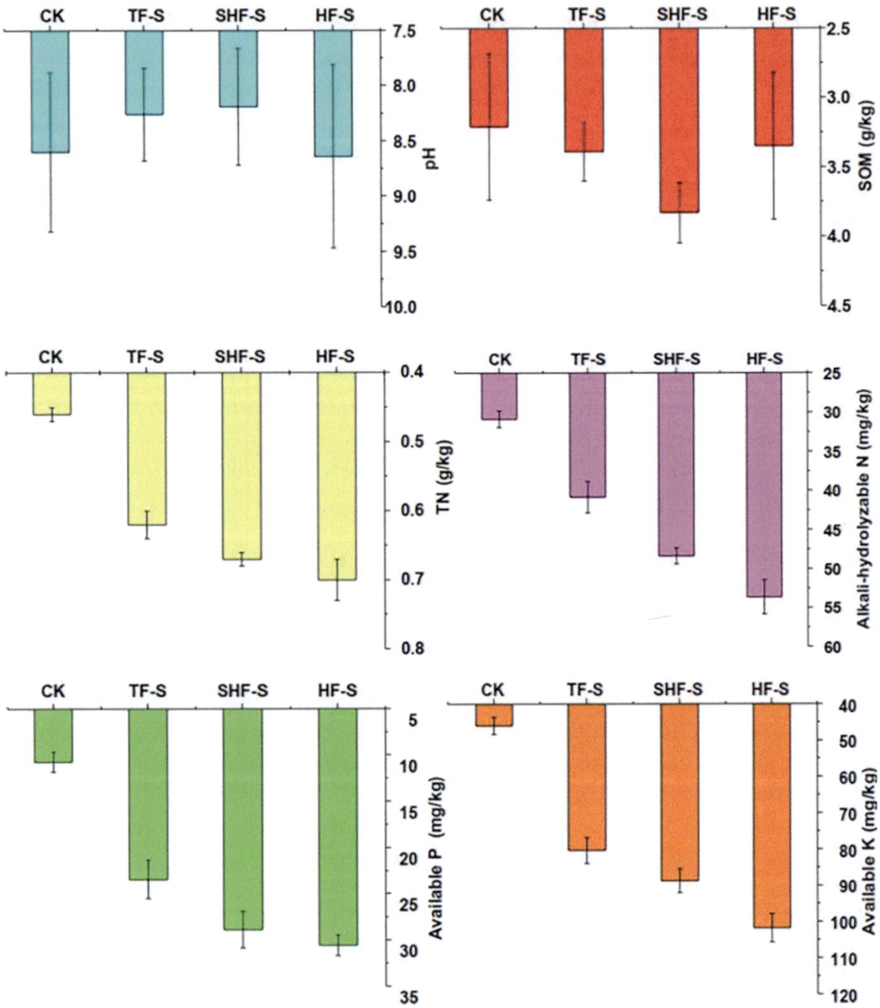

Fig. 4.1 Changes of soil pH and available nutrient under different treatment

Table 4.2 The mass percentage (%) of aggregate size classes in soil with applied exogenous nutrients

Aggregate size class (mm)	CK0	CK-S	TF-S	SHF-S	HF-S
> 2	37.61 ± 4.21	38.38 ± 4.11	31.95 ± 4.17	24.86 ± 2.29	34.44 ± 3.11
1–2	13.57 ± 1.27	12.11 ± 1.22	21.26 ± 2.35	20.58 ± 1.98	23.40 ± 2.06
0.5–1	17.81 ± 2.22	15.55 ± 1.41	22.21 ± 2.67	24.11 ± 2.83	21.3 ± 1.71
0.25–0.5	8.44 ± 0.79	9.21 ± 0.91	8.55 ± 0.64	13.26 ± 2.44	8.13 ± 0.68
0.05–0.25	4.31 ± 0.18	3.17 ± 0.08	5.27 ± 0.10	8.91 ± 0.27	3.66 ± 0.86
< 0.05	18.26 ± 2.35	21.58 ± 3.49	10.76 ± 1.55	8.28 ± 0.91	9.07 ± 0.88

can effectively prevent nutrient loss. The aggregate size distribution in HF-S shows that the application of chemical fertilizer increased the percentage of 0.5–2 mm size fraction aggregates in the soil, but the content of other size fraction aggregates was significantly reduced. The main reason for the above results is that the macromolecules in the food waste and rice husks of TF and SHF, such as proteins, fats, and cellulose, are degraded into large quantities of polysaccharides, reducing sugars, and amino acids during the fermentation and decomposition process. When applied to the soil, these substances have a strong soil particle aggregation capacity, which is conducive to the formation of water-stable aggregates. It has been reported that a good soil structure must ensure that aggregates have a certain stability, especially water stability (Bottinelli et al. 2017).

4.1.2.2 Distribution of Total Organic Carbon in Soil Aggregates

The polymerization and binding action of organic carbon in the soil are key to the formation and stability of soil aggregates; conversely, the granulation process of soil aggregates is also one of the important pathways for soil carbon sequestration (Christensen 1992). Therefore, studying the distribution of organic carbon in the various size fractions of soil aggregates is an important means of understanding the dynamic changes in soil organic matter.

Due to differences in the decomposition rates and residual amounts of different exogenous nutrients in the soil, the application of different exogenous nutrients has a great impact on the TOC content in the various size fractions of soil aggregates. As shown in Table 4.3, overall, the total organic carbon content was lowest in the < 0.05 mm soil aggregate size fraction.

Among the aggregates with a size fraction > 0.25 mm, the 0.25–0.5 mm soil aggregates had the highest organic carbon content, while for the remaining size fractions, the total organic carbon content decreased as the diameter increased. This is because most of the organic carbon in the soil is in a non-labile state and usually uses smaller particulate aggregates as adsorption carriers to form organic–inorganic complex substances. The smaller the diameter of the soil aggregate particles, the larger the specific surface area, and thus the more non-labile carbon is attached to

Table 4.3 TOC in each aggregate class under different treatment (g kg^{-1})

Aggregate size class (mm)	CK0	CK-S	TF-S	SHF-S	HF-S
> 2	4.17 ± 0.29	4.04 ± 0.16	4.51 ± 0.69	4.81 ± 0.94	4.63 ± 0.94
1–2	4.18 ± 0.38	4.24 ± 0.71	4.84 ± 0.44	4.92 ± 0.26	4.71 ± 0.33
0.5–1	4.66 ± 2.14	4.49 ± 0.8	4.87 ± 0.47	5.05 ± 0.10	4.77 ± 0.65
0.25–0.5	4.68 ± 0.22	4.41 ± 0.59	4.96 ± 0.19	5.36 ± 0.67	4.96 ± 0.49
0.05–0.25	4.37 ± 0.51	4.35 ± 0.62	4.71 ± 0.28	4.50 ± 0.25	4.40 ± 0.99
< 0.05	3.16 ± 0.84	3.72 ± 0.26	3.40 ± 0.77	3.44 ± 0.95	3.23 ± 0.12

them. Furthermore, in the initial stage of aggregate formation, the role of organic carbon is more significant than in the later condensation stage of large aggregates. Therefore, the organic carbon content in small-diameter aggregates is higher than that in large-diameter aggregates. Research has shown that soil organic carbon is mainly enriched in 0.25–2 mm aggregates, which is consistent with the results of this study (Cambardella and Elliott 1993; Gupta and Germida 1988).

As an important binding material for aggregates, TOC content is positively correlated with the stability of water-stable aggregates in the soil. Compared to CK-S, the growth rates of TOC content in the various aggregate size fractions for TF-S, SHF-S, and HF-S ranged from 5.60%–14.15%, 3.44%–21.54%, and 0.31%–14.60%, respectively. This indicates that the application of exogenous nutrients can effectively increase the TOC content of each aggregate size fraction in the soil, and that different exogenous nutrients affect aggregate stability by influencing the TOC content in the soil. The application of the SHF conditioner is beneficial for promoting an increase in the total organic carbon content of the 0.25–2 mm aggregate size fraction in the soil. This is because the humic acid in this part of the aggregate size fraction has a relatively high stability, which is conducive to soil carbon sequestration. In the 0.25–1 mm size fraction, the TOC content of SHF-S was the highest, followed by TF-S, HF-S, CK0, and CK-S. This indicates that the unfertilized soil experienced significant mineralization and loss of organic carbon over the long term without the application of exogenous nutrients. In comparison, the application of SHF is more conducive to soil carbon sequestration and promoting enhanced aggregate stability.

4.1.2.3 Distribution of Total Nitrogen in Soil Aggregates

The changes in total nitrogen content in the various size fractions of soil aggregates are similar to those of total organic carbon. As shown in Table 4.4, for aggregates with a size fraction > 0.25 mm, the total nitrogen content gradually increased as the sieve diameter decreased, with the 0.25–0.5 mm size fraction having the highest total nitrogen content and the < 0.05 mm size fraction having the lowest. This is mainly because the fine silt aggregates in the soil are highly susceptible to wind erosion, and the ammonium ions attached to them are easily migrated and transformed. In contrast, aggregates with a size fraction > 0.25 mm act as large water-stable aggregates with high stability and contribute more to soil nitrogen fixation (Lu et al. 2016; Zhao et al. 2017). Secondly, the smaller the aggregate size fraction, the larger the specific surface area and the stronger the adsorption capacity for nitrogen. As the size fraction increases, the specific surface area decreases, and the adsorption of ammonium ions weakens. Therefore, the total nitrogen content in the smaller size fraction aggregates is significantly higher than that in the larger size fraction aggregates.

In the aggregates of soil to which exogenous nutrients were applied, the total nitrogen content for the same size fraction ranked in the order of SHF-S, HF-S, and TF-S. Compared to CK-S, the growth rate ranges for total nitrogen content in each soil aggregate size fraction for TF-S, SHF-S, and HF-S were 33.03%–35.16%, 80.18%–83.67%, and 64.02%–67.03%, respectively. This indicates that the application of

Table 4.4 TN in each aggregate class under different treatment (g kg^{-1})

Aggregate size class (mm)	CK0	CK-S	TF-S	SHF-S	HF-S
> 2	6.51 ± 0.22	6.48 ± 0.65	8.70 ± 0.03	11.72 ± 0.19	10.68 ± 0.14
1–2	6.22 ± 0.36	6.49 ± 0.97	8.73 ± 0.14	11.82 ± 0.77	10.71 ± 0.67
0.5–1	6.48 ± 0.55	6.52 ± 0.31	8.78 ± 0.66	11.84 ± 0.36	10.73 ± 0.61
0.25–0.5	6.51 ± 0.07	6.56 ± 0.19	8.81 ± 0.81	11.84 ± 0.94	10.76 ± 0.32
0.05–0.25	6.39 ± 0.10	6.51 ± 0.07	8.66 ± 0.16	11.73 ± 0.42	10.76 ± 0.54
< 0.05	6.24 ± 0.22	6.37 ± 0.44	8.61 ± 1.04	11.70 ± 0.26	10.64 ± 0.66

exogenous nutrients can increase the total nitrogen content of soil aggregates, with TF and SHF contributing more to the increase in the 0.5–2 mm size fraction, and HF contributing more to the 0.05–0.25 mm size fraction.

4.1.2.4 Distribution of Total Phosphorus in Soil Aggregates

As shown in Table 4.5, under different application methods of exogenous nutrients, the total phosphorus content in each soil aggregate size fraction ranked from high to low as follows: SHF-S, HF-S, TF-S. In aggregates with a size fraction greater than 0.25 mm, as the soil particle size gradually increased, the total soil phosphorus content decreased. The < 0.05 mm size fraction aggregates had the lowest total phosphorus content. Compared to CK0, the change in total phosphorus in CK-S was not significant. This is mainly because the migration rate of phosphorus in soil is low, and typically, soil phosphorus is not lost in a gaseous form (Rowe et al. 2015; Wang et al. 2016c). The contribution of different exogenous nutrient applications to soil total phosphorus content was mainly concentrated in the 0.05–2 mm size fraction aggregates. Among them, TF contributed more to the increase in total phosphorus content in the 0.5–2 mm size fraction, while SHF and HF contributed more to the 0.25–2 mm size fraction. Compared to CK-S, after the application of TF, SHF, and HF, the increase ranges for total phosphorus content in each soil aggregate size fraction were 33.33%–47.06%, 35.09%–54.9%, and 40.35%-56.86%, respectively. From this, it can be seen that HF had the greatest impact on the total phosphorus content of all soil aggregate size fractions, followed by SHF. TF caused the phosphorus in the soil to be primarily stored in the 0.5–2 mm size fraction aggregates, while SHF and HF caused it to be mainly stored in the 0.25–2 mm size fraction aggregates.

Table 4.5 TP in each aggregate class under different treatment (g kg^{-1})

Aggregate size class (mm)	CK0	CK-S	TF-S	SHF-S	HF-S
> 2	0.52 ± 0.03	0.50 ± 0.06	0.75 ± 0.02	0.77 ± 0.02	0.81 ± 0.06
1–2	0.54 ± 0.01	0.51 ± 0.05	0.75 ± 0.10	0.79 ± 0.09	0.80 ± 0.11
0.5–1	0.54 ± 0.01	0.51 ± 0.05	0.75 ± 0.10	0.79 ± 0.09	0.80 ± 0.11
0.25–0.5	0.63 ± 0.03	0.58 ± 0.01	0.78 ± 0.06	0.83 ± 0.08	0.83 ± 0.07
0.05–0.25	0.54 ± 0.01	0.57 ± 0.01	0.76 ± 0.08	0.77 ± 0.10	0.80 ± 0.07
< 0.05	0.50 ± 0.06	0.53 ± 0.01	0.74 ± 0.10	0.72 ± 0.02	0.76 ± 0.09

4.1.3 Characteristics of Precursor Components for Humus Formation

Nitrogenous organic compounds constitute over 90% of total soil nitrogen, with amino acid nitrogen accounting for 30–50% of organic nitrogen. As a key nitrogen source, amino acid nitrogen serves as a vital nutrient for soil microorganisms and plant growth while playing a fundamental role in humus formation. Carbohydrates comprise 5–25% of soil organic matter and represent the primary carbon and energy source for soil microbiota. Notably, polysaccharides and reducing sugars significantly influence the formation and stability of soil macroaggregates through metal ion complexation, acting as precursors for humus synthesis. Carboxyl and phenolic hydroxyl groups are also critical to humification: carboxyl groups attached to humic acid structures facilitate pollutant degradation, while both functional groups enhance electron transfer capacity in organic matter formation. This section examines the dynamics of amino acids, polysaccharides, reducing sugars, carboxyl groups, and phenolic hydroxyl groups during humus transformation following exogenous nutrient amendment.

4.1.3.1 Amino Acids

As depicted in Fig. 4.2, exogenous nutrient amendments markedly increased amino acid content in impoverished soils, with levels ranked as HF-S > SHF-S > TF-S. In unamended barren soil (CK-S), indigenous microbial communities remained subactivated with limited metabolic activity. The addition of exogenous nutrients introduced bioavailable organic compounds, stimulating microbial metabolism. Proteins—essential electron carriers during microbial extracellular respiration—underwent degradation during electron transfer. Rapid and recurrent microbial metabolic cycles led to substantial accumulation of amino acids and microbial residues in later stages, thereby elevating soil amino acid content.

No significant change occurred in CK-S amino acid levels throughout the experiment, indicating minimal impact of nitrogen mineralization, nitrification, or

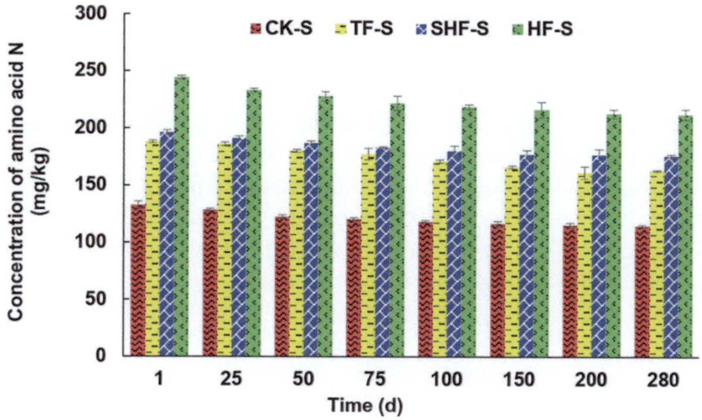

Fig. 4.2 Dynamic changes of amino acid in soil under different treatment

ammonification on nitrogen transformation without exogenous nutrients or crop cultivation.

During days 1–75, amino acids declined gradually across all treatments, stabilizing after day 75. By the experiment's conclusion, TF-S, SHF-S, and HF-S exhibited increases of 42.58%, 53.92%, and 85.51%, respectively. However, compared to initial values post-amendment, net reductions of 13.16% (TF-S), 10.29% (SHF-S), and 13.37% (HF-S) were observed. This demonstrates that exogenous nutrients enhance amino acid availability in nutrient-deprived soils, where microbial diversity and abundance are initially suppressed. Following amendment, amino acids were rapidly metabolized by microbes, synthesized, and stored to support sustained nitrogen supply for crops.

4.1.3.2 Polysaccharides and Reducing Sugars

Soil polysaccharide content is influenced by tillage practices, land use patterns, and plant species. Polysaccharides constitute 6–10% of soil organic carbon and critically regulate humus formation and stability. Figure 4.3 shows pronounced treatment effects on polysaccharide levels in barren soil: SHF-S > TF-S > HF-S. CK-S exhibited negligible variation. TF and SHF amendments significantly increased polysaccharides; at day 280, TF-S and SHF-S reached 89.79 mg g^{-1} and 117.37 mg g^{-1}, respectively—27.78% and 36.62% higher than CK-S. This aligns with studies linking polysaccharide accumulation to microbial biomass, as organic fertilizers enhance microbe-mediated aggregation of soil particles into water-stable aggregates. Conversely, HF-S decreased polysaccharides by 0.22 mg g^{-1} relative to CK-S. Polysaccharide content correlates positively with organic fertilizer dosage but negatively with inorganic nitrogen. Chemical fertilizers induce slight soil pH elevation, which modulates microbial activity and nutrient availability, thereby affecting polysaccharide synthesis.

Reducing sugar dynamics are shown in Fig. 4.4. CK-S levels remained stable, while HF-S declined from 0.04 mg g^{-1} (day 1) to 0.034 mg g^{-1} (day 280). In contrast, TF and SHF treatments induced initial increases peaking at 0.18 mg g^{-1} (TF-S) and 0.22 mg g^{-1} (SHF-S) on day 100, followed by declines. By day 280, TF-S and SHF-S still exceeded CK-S by 0.13 mg g^{-1} and 0.14 mg g^{-1}, confirming their strong stimulatory effect on reducing sugars.

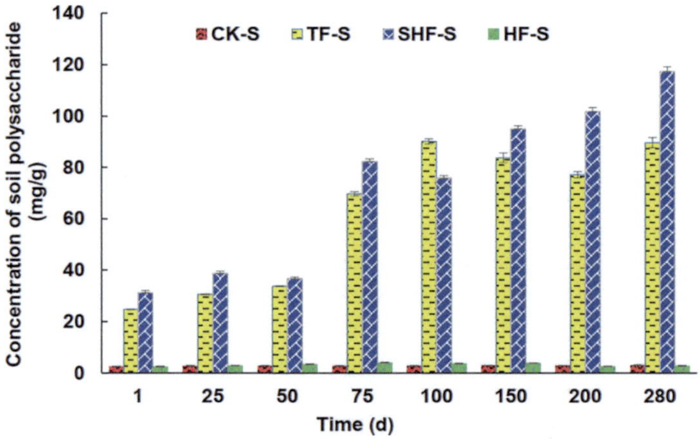

Fig. 4.3 Dynamic changes of polysaccharide in soil under different treatment

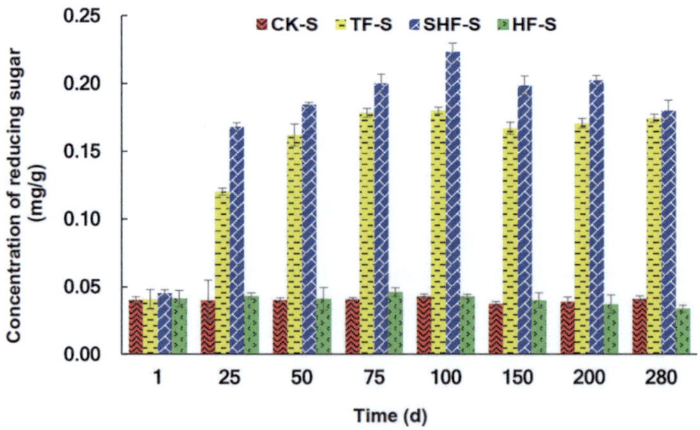

Fig. 4.4 Dynamic of reducing sugar in soil under different treatment

4.1.3.3 Carboxyl and Phenolic Hydroxyl Groups

As illustrated in Fig. 4.5, CK-S carboxyl concentrations showed minor reduction without statistical significance. Exogenous nutrients significantly altered carboxyl profiles: HF-S increased moderately to 8.36 mmol g^{-1} (6.05 mmol g^{-1} above CK-S) by day 280, attributable to NPK-enhanced microbial production of humus precursors. TF-S and SHF-S exhibited substantial rises, exceeding CK-S by 7.46 mmol g^{-1} and 10.36 mmol g^{-1}, respectively. This stems from protein degradation in food waste-derived amendments (TF/SHF), generating nitrogenous compounds that form carboxyl groups during fermentation. These carboxyl groups integrate into humic macromolecules via binding to core structural units, elevating detectable soil concentrations.

Phenolic hydroxyl groups followed similar trends (Fig. 4.6). CK-S remained stable, while amendments sharply increased levels—TF-S, SHF-S, and HF-S surpassed CK-S by 319.48%, 414.29%, and 164.03%, respectively. This corroborates Galantini and Rosell (2006), who reported organic fertilizer-induced increases in humic acid phenolic groups. The pronounced effect of TF/SHF arises from lignin and cellulose degradation during fermentation, releasing phenolic hydroxyl precursors. Studies confirm that microbially transformed phenolic compounds condense with amino acids to form humus. Additionally, lignin in TF/SHF and native soil undergoes three pathways: (1) degradation to phenols → quinones; (2) direct quinone formation as humic acid building blocks; (3) microbial recalcitrance to form humic acid cores that cross-link with small molecules—collectively enhancing humification in barren soils.

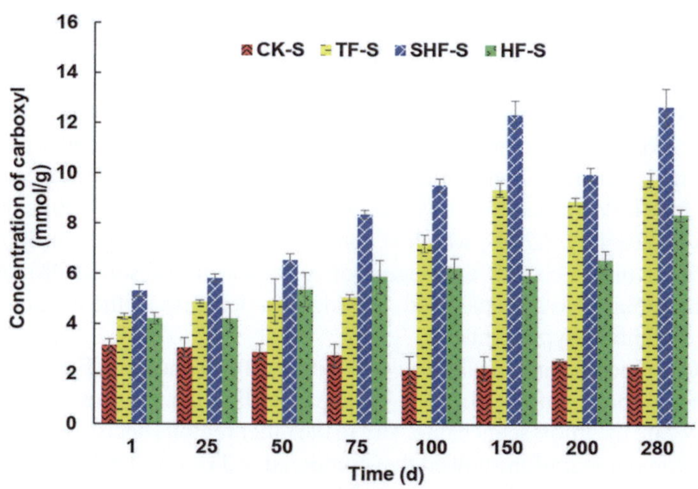

Fig. 4.5 Concentration of carboxyl in soil under different treatment

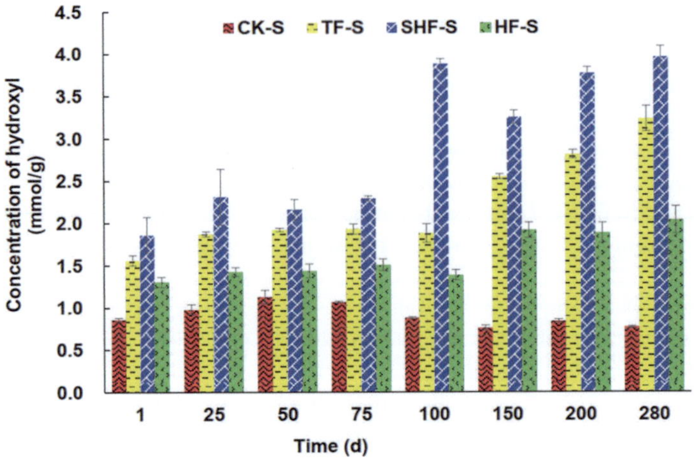

Fig. 4.6 Concentration of hydroxyl in soil under different treatment

4.1.4 Fluorescence Spectroscopy Analysis of Soil Humic Substances

4.1.4.1 Three-Dimensional Fluorescence Spectroscopy of Soil DOM

The fluorescence spectrum is divided into five regions, I-V, with the components corresponding to the *Pi,n* value of each region being the same as in Sect. 2.2.2.1. As shown in Fig. 4.7, the fluorescence intensity changes in each region of CK-S were not significant. The fluorescence signal intensity in regions III and V increased slightly as the treatment time was extended, while the fluorescence signal intensity in region IV weakened slightly. From the volumetric integration ratio of the DOM three-dimensional fluorescence regions, it is evident that the *Pi,n* in regions III and V of CK-S was significantly higher than in other regions. At the end of the treatment, the *Pi,n* of region III decreased by 2.63% compared to the initial value, while the *Pi,n* of region V increased by 2.29%, which is consistent with the changes in fluorescence intensity. This indicates that the degree of humification of soil DOM increased, and soil microbial activity weakened, leading to a decrease in the accumulation of microbial metabolic byproducts.

The DOM fluorescence peak scan intensity of TF-S showed (Fig. 4.8) that the fluorescence signal intensity in region V increased slightly as the treatment time was extended, while the fluorescence intensity in the other regions decreased slightly. The regional volume integration showed that compared to TF-S-1, the *Pi,n* in region V of the DOM structure of TF-S-280 increased by 9.22%, while the *Pi,n* in regions I, II, III, and IV decreased by 6.58%, 3.24%, 10.43%, and 24.83%, respectively, further verifying the conclusions drawn from the fluorescence peak intensity experiment.

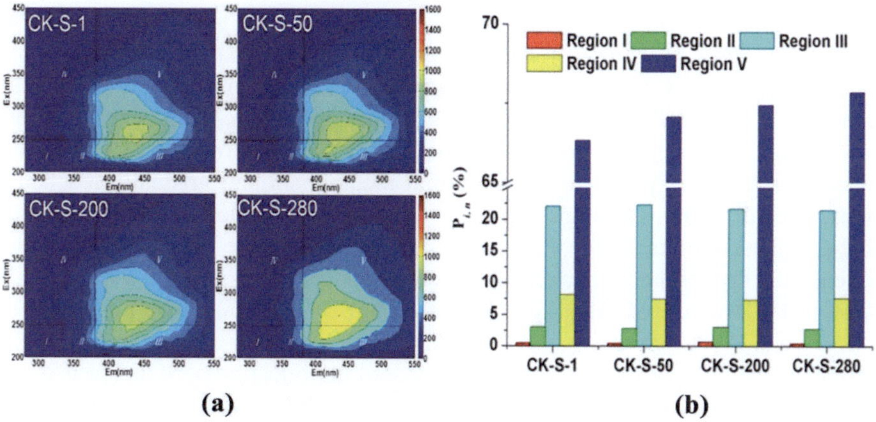

Fig. 4.7 Excitation–emission matrix spectra (**a**) and volumetric fluorescence distribution of DOM extracted from original soil (CK-S) (**b**). CK-S-1, CK-S-50, CK-S-200, and CK-S-280 represent the CK treatment for 1 d, 50 d, 200 d, and 280 d

This indicates that the application of TF promoted the formation of soil humus and was beneficial for increasing the degree of soil humification.

The DOM fluorescence peak scan intensity of the SHF-S treatment group showed that its structure underwent significant changes (Fig. 4.9). The three-dimensional fluorescence spectrum showed that, overall, the degree of humification of SHF-S was lower than that of TF-S. Compared to the first day of treatment, the fluorescence signal intensity in region V was significantly enhanced on the 200th day of treatment. In the following 80 days, the rate of increase of the DOM fluorescence signal intensity slowed down. This indicates that the effect of SHF application on promoting soil humification was significant in the initial stage. In the later stage of application, the

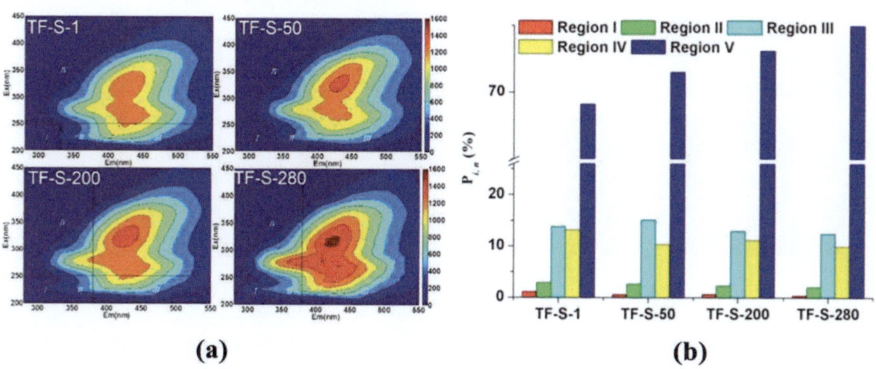

Fig. 4.8 Excitation–emission matrix spectra (**a**) and volumetric fluorescence distribution of DOM extracted from TF-S (**b**). TF-S-1, TF-S-50, TF-S-200, and TF-S-280 represent the soil with TF for 1 d, 50 d, 200 d, and 280 d

rate of soil humus synthesis gradually slowed. This is related to the fact that SHF is rich in a large amount of humus synthesis precursors. In the initial stage of application, under the action of soil microorganisms, the humus precursors actively participate in the synthesis of soil humus. In the later stage of application, the consumption rate of SHF conditioner humus precursors was lower than the formation rate of small-molecule substances produced by microbial metabolism and gradually reached a dynamic equilibrium, causing the humus synthesis rate to slow down. The results of the regional volume integration were consistent with the results of the three-dimensional fluorescence spectroscopy. Compared to SHF-S-1, the *Pi,n* in region V of the DOM structure of SHF-S-280 increased by 12.74%, while the *Pi,n* in regions I, II, III, and IV decreased by 47.17%, 20.37%, 28.67%, and 35.11%, respectively. Compared to TF-S, although the overall degree of humification of SHF-S was lower within the experimental period, the growth rate of its fluorescence region V volume integration was 3.52% higher than that of TF-S. This indicates that a certain amount of undecomposed organic material in SHF was transferred from above ground to the soil for composting, which promoted the activation of old humus, released nutrients, and participated in the formation of microaggregates. Therefore, SHF promotes the degree of soil humification at a faster rate.

The DOM fluorescence signal intensity of the HF-S treatment group showed a similar trend of change to that of TF-S and SHF-S. The regional volume integration showed (Fig. 4.10) that compared to the initial values, the *Pi,n* in regions I-IV of the DOM structure decreased by 55.56%, 46.82%, 26.27%, and 17.43%, respectively. Conversely, the *Pi,n* in region V increased by 7.5%. In summary, the application of different exogenous nutrients has a certain promoting effect on the humification of soil DOM structure. The degree of promotion of soil DOM humification, from high to low, is in the order of SHF > TF > HF.

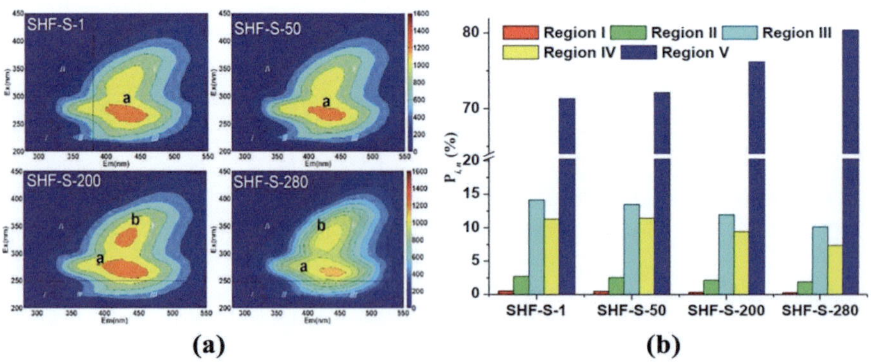

Fig. 4.9 Excitation–emission matrix spectra (**a**) and volumetric fluorescence distribution of DOM extracted from SHF-S (**b**). SHF-S-1, SHF-S-50, SHF-S-200, and SHF-S-280 represent the soil with SHF for 1 d, 50 d, 200 d, and 280 d

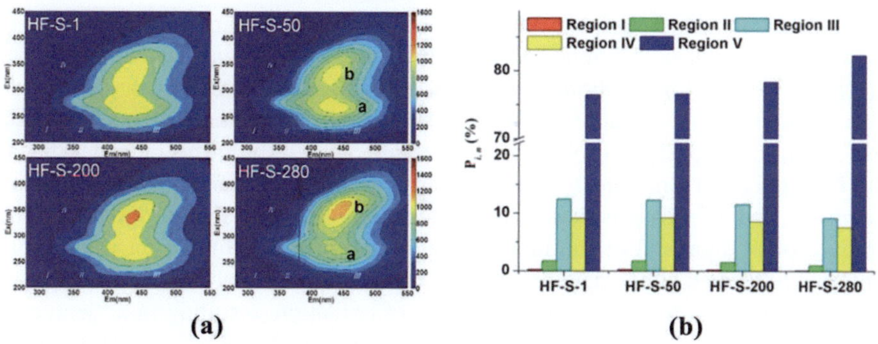

Fig. 4.10 Excitation–emission matrix spectra (**a**) and volumetric fluorescence distribution of DOM extracted from HF-S (**b**). HF-S-1, HF-S-50, HF-S-200, and HF-S-280 represent the soil with HF for 1 d, 50 d, 200 d, and 280 d

4.1.4.2 Three-Dimensional Fluorescence Spectroscopy of Soil Humic Acid

Soil humic acid is a major component of soil humus and plays a crucial role in the formation of soil structure. The 3D-EEM spectrum of soil humic acid shows (Fig. 4.11a) that in unfertilized soil, similar to the structural changes in soil DOM, the fluorescence intensities of regions IV and V in the soil humic acid structure are consistently dominant. The proportions occupied by the volumetric integrals of different fluorescence regions show (Fig. 4.6b) that compared to CK-S-1, the *Pi,n* of regions III and V decreased by 4.55% and 10.84%, respectively. Conversely, the *Pi,n* of regions I, II, and IV increased by 12.79%, 30.20%, and 30.98%, respectively.

The 3D-EEM spectrum of soil humic acid from the TF-S treatment group shows two fluorescence peaks (Fig. 4.12a), corresponding to protein-like substances and

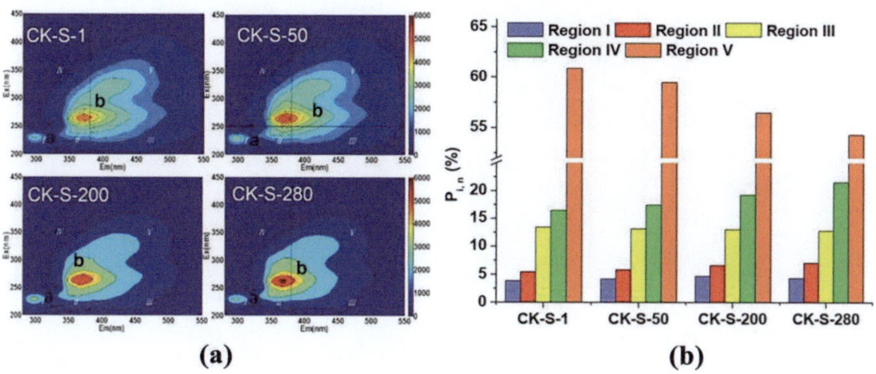

Fig. 4.11 Excitation–emission matrix spectra (**a**) and volumetric fluorescence distribution of HA from the original soil (**b**). CK-S-1, CK-S-50, CK-S-200, and CK-S-280 represent the CK soil treatment group for 1 d, 50 d, 200 d, and 280 d

humic-like substances, respectively. As the duration of TF application increased, the fluorescence peak "a", representing protein-like substances, became a shoulder peak and its signal intensity decreased, disappearing by day 280 of the treatment. The signal intensity of fluorescence peak "b", representing humic-like substances, increased, and its peak position exhibited a significant red shift. This indicates that as the TF application time increased, the degree of aromatization and polycondensation increased, and TF was beneficial for enhancing the degree of soil humification. The application of organic fertilizer leads to the formation of new humic acid in the soil. This newly formed humic acid has a relatively small molecular weight, a decreased degree of condensation, and a lower degree of oxidation, resulting in a "younger" molecular structure (Dou 2010). The fluorescence regional volume integration shows (Fig. 4.12b) that compared to TF-S-1, the *Pi,n* of regions I-IV in TF-S-280 decreased by 93.22%, 84.16%, 47.52%, and 63.0%, respectively, while the *Pi,n* of region V increased by 23.94%. The application of TF helps to promote the formation of humic-like substances in the soil.

The 3D-EEM spectrum of the soil humic acid structure from the SHF-S treatment group shows a total of three fluorescence peaks, "a", "b", and "c" (Fig. 4.13a). As the treatment time increased, peak "a" disappeared, the fluorescence intensity of peak "b" increased significantly, and its peak position showed a clear red shift, indicating that the application of SHF is conducive to the degradation of protein-like substances in the soil. With the application of SHF, the molecular weight of hydrophilic components increased, and the degree of soil humification improved. In the SHF-S-280 treatment group, peak "c" appeared, indicating the accumulation of fulvic-like substances. Research has shown that the application of conditioners tends to make newly formed humus more fulvic acid-like (Chiu et al. 2016). The fluorescence regional volume integration shows (Fig. 4.13b) that compared to the first day of SHF application, on the 280th day, the *Pi,n* of soil humic acid regions

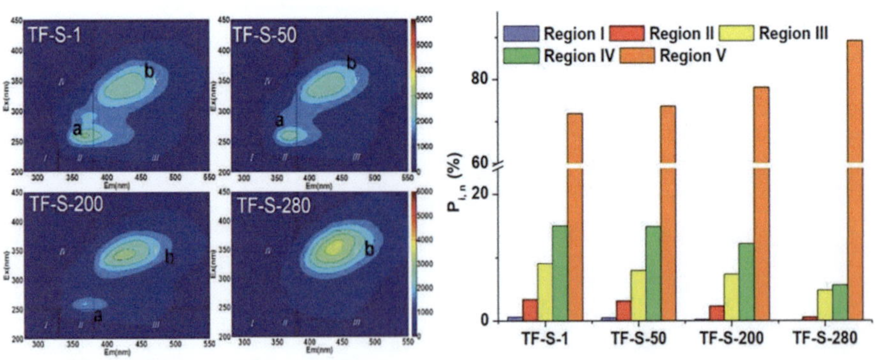

Fig. 4.12 Excitation–emission matrix spectra (**a**) and volumetric fluorescence distribution of HA from the TF-S (**b**). TF-S-1, TF-S-50, TF-S-200, and TF-S-280 represent the soil with TF for 1 d, 50 d, 200 d, and 280 d

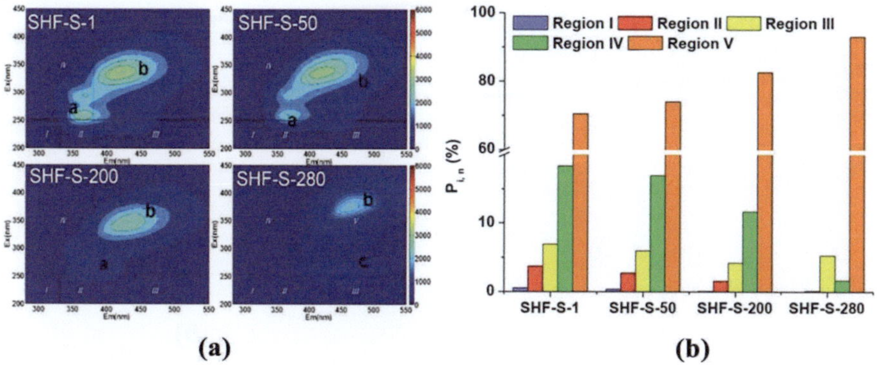

Fig. 4.13 Excitation–emission matrix spectra (**a**) and volumetric fluorescence distribution of HA from the SHF-S (**b**). SHF-S-1, SHF-S-50, SHF-S-200, and SHF-S-280 represent the SHF-S treatment group for 1 d, 50 d, 200 d, and 280 d

I-IV had decreased by 100%, 96.76%, 23.58%, and 91.05%, respectively, while the $P_{i,n}$ of region V had increased by 31.84%. This indicates that the application of SHF promoted the formation of humic-like substances in the soil.

The 3D-EEM spectrum of the HF-S humic acid structure shows two fluorescence peaks, "a" and "b" (Fig. 4.14a). As the HF application time increased, the fluorescence signal intensity of peak "a" decreased significantly, indicating that HF aids in the degradation of protein-like substances in the soil. The fluorescence signal intensity of peak "b" showed a slight increase. The fluorescence regional volume integration shows (Fig. 4.14b) that compared to the first day of application, after 280 days of HF application, the $P_{i,n}$ of humic acid structure regions I-IV decreased by 77.38%, 48.33%, 26.36%, and 31.56%, respectively, while the $P_{i,n}$ of region V increased by 15.48%. The application of HF helps to promote the formation of humic-like substances in the soil. In summary, the intensity of the increase in soil humic acid due to the application of exogenous nutrients is in the order of SHF > TF > HF, which is consistent with the results for the DOM increase.

4.1.4.3 Three-Dimensional Fluorescence Spectroscopy of Soil Fulvic Acid

Fulvic acid is an important component of soil humus. It has a small relative molecular weight, high activity, and a high degree of oxidation. It can participate in nitrogen fixation, significantly increasing the bioavailability of newly fixed ammonium nitrogen, and is closely related to the migration of nutrients in the soil. It is also more sensitive to the application of organic materials than humic acid. Furthermore, due to its smaller molecular weight and higher carboxyl content, fulvic acid has a strong ion chelation capacity and also plays an important role in promoting plant root growth, increasing the number and length of secondary roots, and improving crop yields.

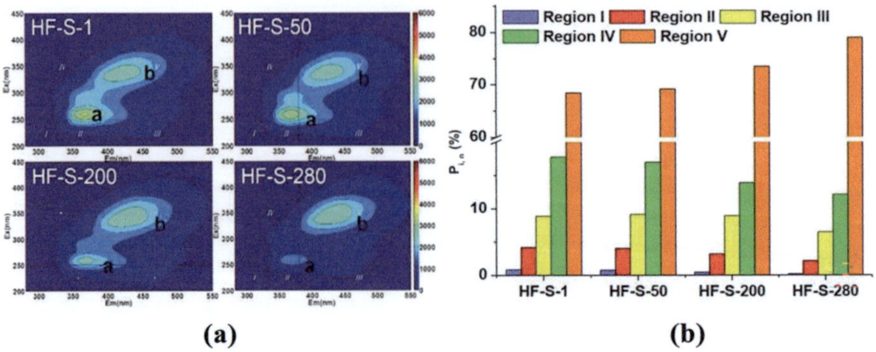

Fig. 4.14 Excitation–emission matrix spectra (**a**) and volumetric fluorescence distribution of HA from the HF-S (**b**). HF-S-1, HF-S-50, HF-S-200, and HF-S-280 represent the soil with HF for 1 d, 50 d, 200 d, and 280 d

The 3D-EEM spectrum of unfertilized soil fulvic acid shows (Fig. 4.15a) that the signal intensity of the fluorescence peak in region V gradually weakened as the treatment time increased. This indicates that the fulvic acid content in the unfertilized soil gradually decreased. The change in regional volume integration shows (Fig. 4.15b) that among the five regions, the *Pi,n* of regions III and V consistently remained at a high level. Compared to the initial value, the *Pi,n* of region V in the fulvic acid structure increased by 6.56% after 200 days of treatment. After 280 days of treatment, the *Pi,n* of fulvic acid structure regions I, III, and V decreased by 29.49%, 5.58%, and 4.07%, respectively, while the *Pi,n* of regions II and IV increased by 58.75% and 43.16%, respectively.

The 3D-EEM spectrum of the fulvic acid structure from the TF-S treatment group shows (Fig. 4.16a) that compared to the first day of TF application, the fluorescence peak signal of the soil fulvic acid structure weakened after 200 days of TF application.

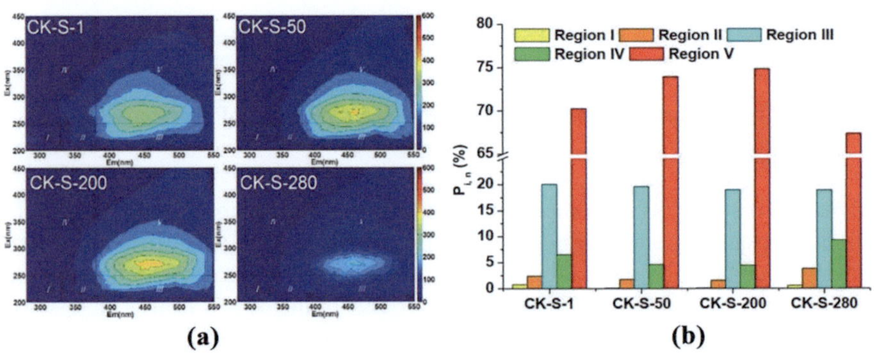

Fig. 4.15 Excitation–emission matrix spectra (**a**) and volumetric fluorescence distribution of FA from the original soil (CK-S) (**b**). CK-S-1, CK-S-50, CK-S-200, and CK-S-280 represent the CK soil treatment group for 1 d, 50 d, 200 d, and 280 d

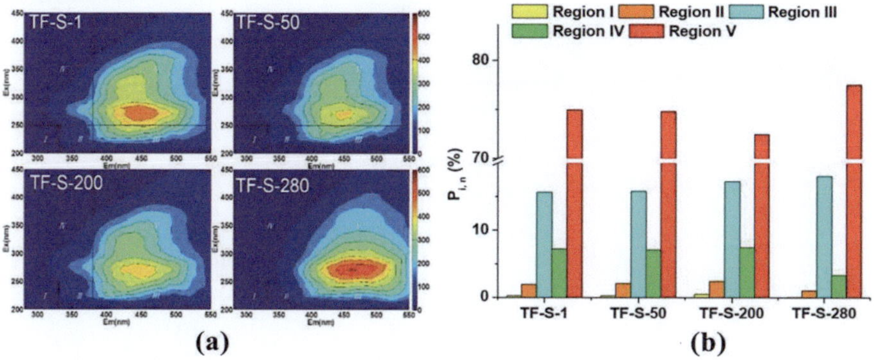

Fig. 4.16 Excitation–emission matrix spectra (**a**) and volumetric fluorescence distribution of FA from the TF-S (**b**). TF-S-1, TF-S-50, TF-S-200, and TF-S-280 represent the soil with TF for 1 d, 50 d, 200 d, and 280 d

On the 280th day of application, the fluorescence peak signal intensity increased significantly, indicating that TF can significantly increase the percentage of soil fulvic acid in the later stages of application. The change in regional volume integration shows (Fig. 4.16b) that on the 200th day of TF application, the *Pi,n* of regions I-IV increased by 85.18%, 24.48%, 10.37%, and 2.49%, respectively, compared to the initial application period, while the *Pi,n* of region V decreased by 3.32%. On the 280th day of TF application, the *Pi,n* of regions I, II, and IV decreased by 74.07%, 45.31%, and 54.43%, respectively, while regions III and V increased by 15.69% and 3.42%, respectively, which is the opposite of the regional volume integration results for the fulvic acid structure of the unfertilized soil. This indicates that TF can effectively promote the formation of soil fulvic acid structures. For some structurally simple fulvic acids, the application of organic fertilizer or organic materials increases the degree of condensation, aromaticity, and the number of aliphatic chain hydrocarbons in the soil fulvic acid, enhancing its effect on soil fertility (Gigliotti et al. 2003; Li et al. 2014a).

The 3D-EEM spectrum of the fulvic acid structure from the SHF-S treatment group shows (Fig. 4.17a) that as the treatment time increased, the fluorescence peak intensity gradually strengthened, indicating that the application of SHF is conducive to the formation of soil fulvic acid structures. The regional volume integration shows (Fig. 4.17b) that compared to the first day of application, after 280 days of SHF application, the integral changes in each region of the arable soil fulvic acid were significant. The *Pi,n* of regions I, II, and IV decreased by 81.33%, 59.49%, and 58.57%, respectively, while regions III and V increased by 4.82% and 12.63%, respectively. This is consistent with the changes in the TF-S treatment group. The increase in the volumetric integral of region V for SHF-S fulvic acid was 9.21% higher than that for TF-S, indicating that compared to traditional fertilizer, the synergistic and high-efficiency conditioner promotes the formation of soil fulvic acid at a faster rate.

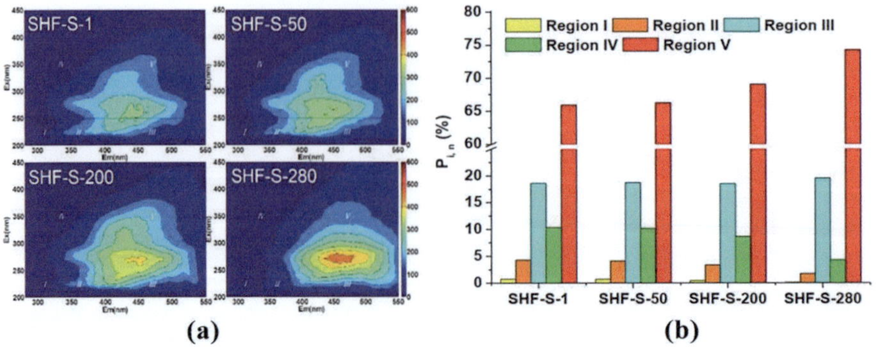

Fig. 4.17 Excitation–emission matrix spectra (**a**) and volumetric fluorescence distribution of FA from the SHF-S (**b**). SHF-S-1, SHF-S-50, SHF-S-200, and SHF-S-280 represent the soil with SHF for 1 d, 50 d, 200 d, and 280 d

The 3D-EEM spectrum of the fulvic acid structure from the HF-S treatment group shows (Fig. 4.18a) that from days 1–200 of HF application, the fluorescence peak signal intensity showed a slow upward trend. On the 280th day of HF application, the fluorescence peak signal intensity decreased slightly. This indicates that HF can rapidly promote the accumulation of soil fulvic acid in the initial stage of application, but its duration is short, and its effect is significantly reduced in the later stage of application (200–280 days). The fulvic acid regional volume integration shows (Fig. 4.18b) that compared to the first day of application, on the 280th day of HF application, the *Pi,n* of regions I-IV decreased by 55.81%, 17.92%, 3.02%, and 7.28%, respectively, while the *Pi,n* of region V increased by 3.0%. In summary, the influence of the three exogenous nutrients on the increase in soil fulvic acid, in order of intensity from high to low, is SHF > TF > HF, which is consistent with the results for the increase in soil humic acid.

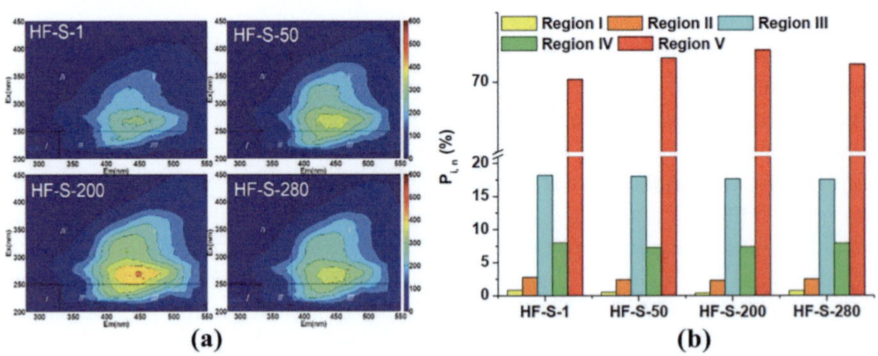

Fig. 4.18 Excitation–emission matrix spectra (**a**) and volumetric fluorescence distribution of FA from the HF-S (**b**). HF-S-1, HF-S-50, HF-S-200, and HF-S-280 represent the soil with HF for 1 d, 50 d, 200 d, and 280 d

4.1.5 ^{13}C-NMR Spectroscopy Analysis of Soil Humic Acid

The characteristic peaks in the ^{13}C-NMR spectra of soil humic acid were characterized using the classic four-region division method reported by Dou et al. (2010). As shown in Fig. 4.19, the ^{13}C-NMR spectrum of humic acid revealed that the tested soil humic acid contains four carbon functional groups: alkyl C (0–50 ppm), O-alkyl C (50–110 ppm), aromatic C (110–160 ppm), and carbonyl C (160–200 ppm). In the CK-S treatment, the maximum absorption in the alkyl C region was at approximately δ 31 ppm, attributed to the contribution of polymethylene chains in saturated hydrocarbons. In the O-alkyl C region, the maximum absorption signal was between δ 56–58 ppm, which is the characteristic absorption range for methoxy C. The absorption signal for secondary alcohol C in carbohydrates appeared between 74 and 76 ppm. The strongest signal peaks in the aromatic C region were at 127–129 ppm and 154 ppm, corresponding to H-substituted aromatic C ortho and para to substituents like O and N, and to phenolic hydroxyl C, respectively. The signal δ in the carbonyl C region was mainly at 173–180 ppm, with a weak signal in the 190–200 ppm area. Therefore, in the CK-S humic acid, carbonyl C from carboxylic acids, esters, and amides was dominant, while the content of carbonyl C from aldehydes and ketones was minimal. During the experiment, the spectral shapes of humic acid in the CK-S 1 d, 50 d, and 280 d treatment groups were fundamentally consistent, but the absorption intensity of the aforementioned chemical shift peaks changed significantly. As shown in Fig. 4.18, at the end of the experiment, the proportion of functional group carbon distribution in the 160–200 ppm absorption region of CK-S humic acid decreased significantly, indicating that the content of carbonyl C in its carboxylic acids, esters, and amides decreased over time.

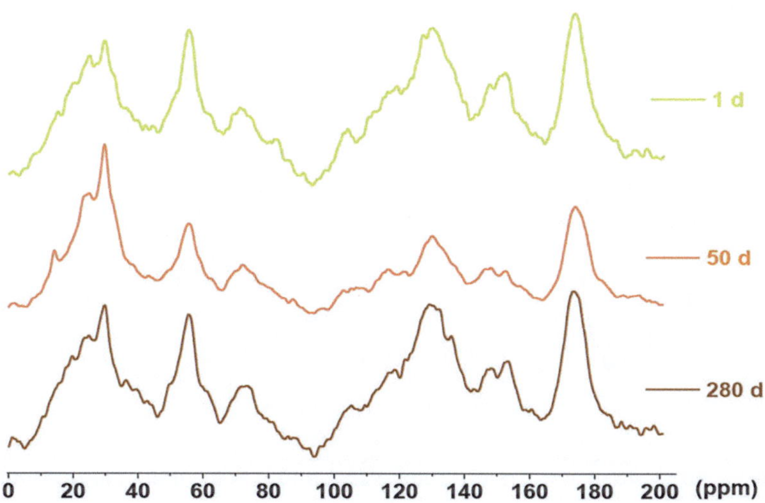

Fig. 4.19 Solid state ^{13}C cross-polarization magic-angle spinning (CPMAS) NMR spectra of humic acids in CK-S

In the TF-S treatment (as shown in Fig. 4.20), the positions of the signal peaks for each functional group were the same as in the CK-S treatment group. As the treatment time increased, the signal peak intensity at 0–50 ppm gradually increased, while the signal peak intensity at 110–160 ppm slightly decreased. Compared to day 1 of TF application, the alkyl C content in the TF-S treatment group decreased by 15.95% on day 200, and rebounded by 23.07% on day 280. This indicates that the application of TF is conducive to increasing the soil alkyl C content and promotes the polycondensation and chain extension of polymethylene chains in the later stages of application. On day 280 of TF application, the proportions of alkyl C, O-alkyl C, and carbonyl C increased by 3.44%, 1.38%, and 6.82%, respectively, compared to day 1 of TF application, while the aromatic C content decreased by 6.83%. Even though the aromatic C content decreased, aromatic C remained the dominant carbon type throughout the treatment periods, indicating that the soil with TF application has a distinct aromatic structure. As shown in Fig. 4.18, at the end of the experiment, the total aliphatic C content (the sum of O-alkyl C and alkyl C) of its humic acid was 50.16%, whereas the total aliphatic C content in the CK-S treatment group was 46.88%. This suggests that the application of TF organic fertilizer led to the aliphatization of humic acid, which will be beneficial for increasing soil humic acid activity.

In the SHF-S treatment group, in addition to the aforementioned chemical shift peaks, a small peak appeared at 104–105 ppm, which is the signal for di-O-alkyl C (acetal) in polysaccharides. This indicates that the application of the SHF soil conditioner is conducive to increasing the soil polysaccharide content. As shown in Fig. 4.21, with increasing treatment time, the content of polymethylene C and aromatic C in the humic acid structure decreased, while the content of methoxy C,

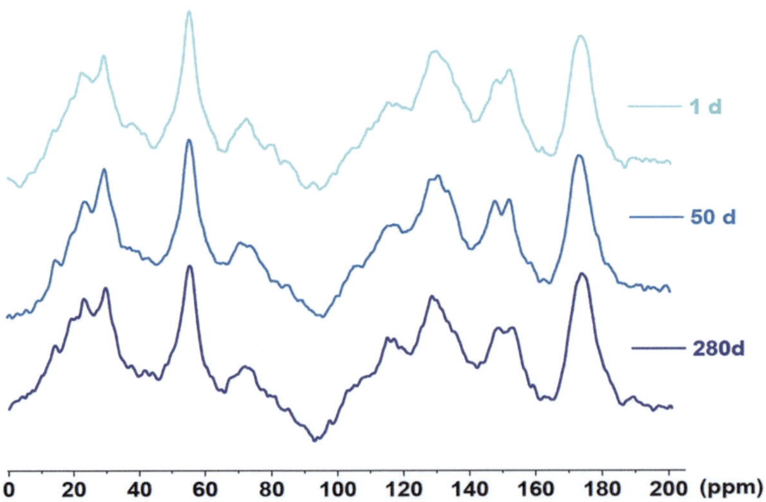

Fig. 4.20 Solid state ^{13}C cross-polarization magic-angle spinning (CPMAS) NMR spectra of humic acids in soil with TF

alcohol hydroxyl C, and carbonyl C increased. This suggests that the application of SHF can increase the oxidation degree of soil humic acid. It can be seen from Fig. 4.23 that although the alkyl C content showed a slight decreasing trend with increasing treatment time, it remained the dominant carbon type in all treatment periods. A high alkyl C content implies an increase in the stability of soil organic carbon. Furthermore, after the application of SHF, the total aliphatic C content of humic acid was 59.94%, which was 13.06% and 9.78% higher than that of the CK-S and TF-S treatment groups, respectively. This demonstrates that the application of the synergistic and high-efficiency soil conditioner enhances the activity of soil humic acid more significantly than the no-fertilizer group and the organic fertilizer group.

In the HF-S treatment, the positions of the signal peaks for each functional group were the same as in the TF-S and SHF-S treatment groups. With increasing treatment time, the signal peak intensity at 0–50 ppm gradually increased, the signal peak intensity in the 110–160 ppm range decreased significantly, and there were no obvious changes in the intensity of other signal peaks. Figure 4.22 shows that compared to day 1 of HF application, the functional carbon content in the 0–110 ppm range of the soil humic acid structure increased by 2.79% on day 50 of HF application, but its content decreased by 8.05% on day 280. This indicates that the application of HF has a significant promoting effect on the increase of methoxy C content in humic acid mainly in the early stage. Compared to day 1 of application, its carbonyl C content decreased by 18.58%, indicating that the application of HF significantly reduced the oxidation degree of soil humic acid. The total aliphatic C content of the humic acid

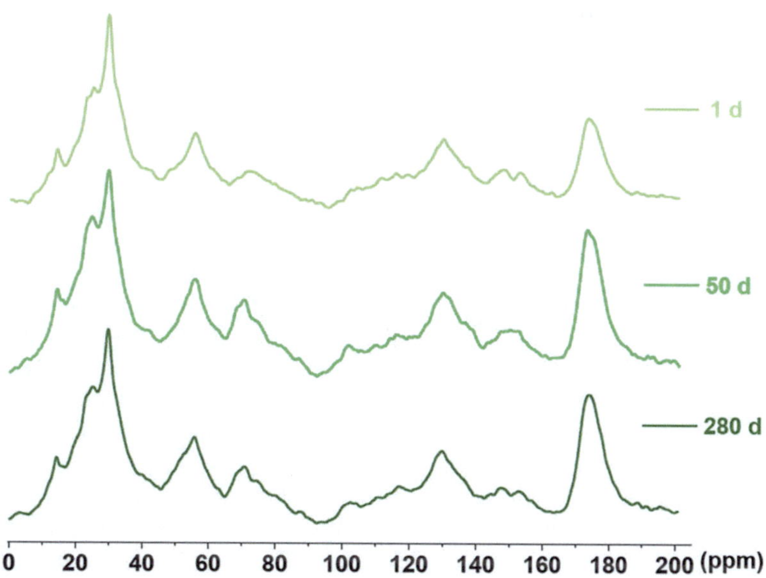

Fig. 4.21 Solid state ^{13}C cross-polarization magic-angle spinning (CPMAS) NMR spectra of humic acids in soil with SHF

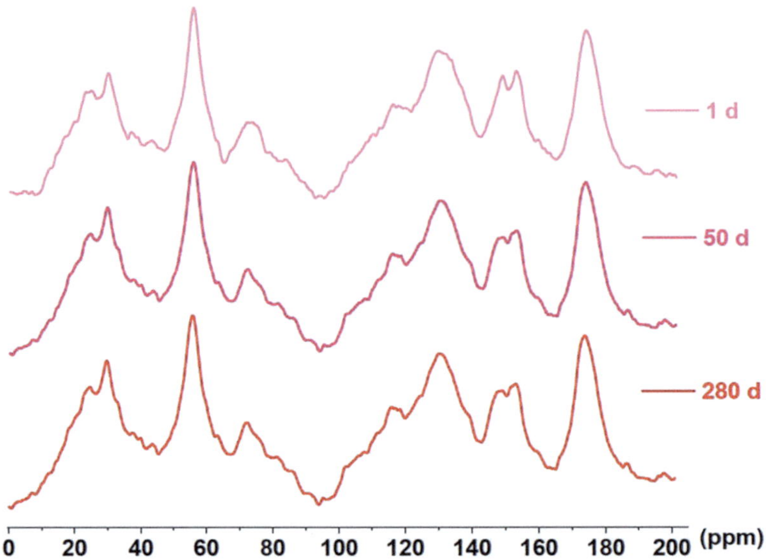

Fig. 4.22 Solid state ^{13}C cross-polarization magic-angle spinning (CPMAS) NMR spectra of humic acids in soil with HF

was 49.9%, an increase of 3.02% over CK-S, suggesting that the application of HF caused the aliphatization of humic acid. In summary, the application of TF, SHF, and HF all improved soil humic acid activity to varying degrees, with the intensity in the order of SHF > TF > HF.

4.1.6 Changes in the Soil Carbon Pool Management Index

As shown in Table 4.6, the application of TF, SHF, and HF had a significant impact on soil TOC, LOC, and NLOC. Compared with CK-S, the application of both SHF and HF increased the contents of soil TOC, LOC, and NLOC to varying degrees. The application of TF increased the contents of soil TOC and LOC, but decreased the NLOC content. The increases in soil LOC content for TF, SHF, and HF were 0.22 g kg^{-1}, 0.31 g kg^{-1}, and 0.07 g kg^{-1}, respectively. A high LOC content in the soil represents a higher degradation rate of soil organic matter and a faster nutrient cycle rate, which can increase the nutrient transformation rate in the soil and promote crop growth (Wang et al. 2015).

The CPMI index can comprehensively describe the amount of soil organic carbon and soil fertility. The CPMI index of SHF-S was 1.09 times and 1.36 times that of the TF-S and HF-S treatment groups, respectively. This indicates that the application of the SHF soil conditioner can increase the accumulation of soil organic carbon

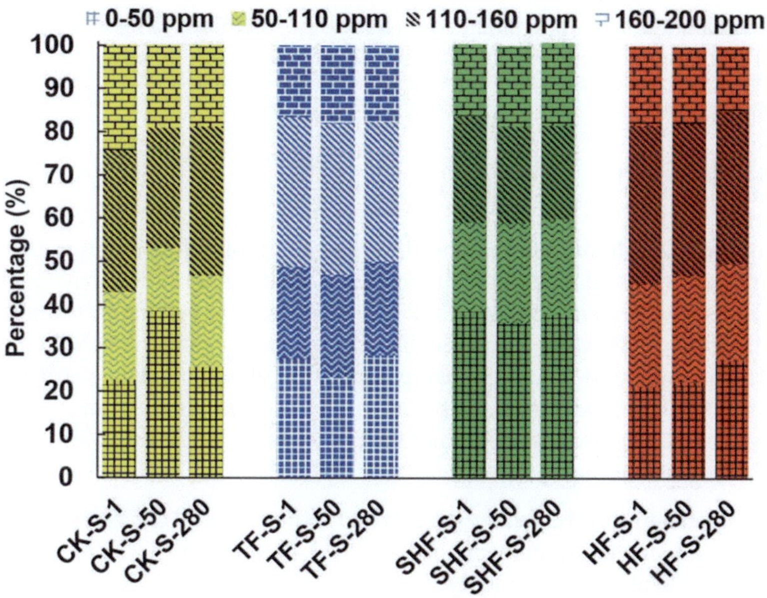

Fig. 4.23 Distribution of C functional groups of soil humic acids in different treatments

Table 4.6 Effects of exogenous nutrients on the CPMI of organic-matter impoverished soil

	TOC	LOC	NLOC	CPI	AI	CPMI
	(g kg^{-1})	(g kg^{-1})	(g kg^{-1})			
CK	3.02 ± 0.02	0.81 ± 0.01	2.21 ± 0.01	1	1	100
TF-S	3.12 ± 0.03	1.03 ± 0.02	2.02 ± 0.03	1.033 ± 0.03	1.345 ± 0.03	138.91 ± 0.03
SHF-S	3.38 ± 0.01	1.12 ± 0.01	2.26 ± 0.01	1.119 ± 0.01	1.352 ± 0.01	151.33 ± 0.01
HF-S	3.10 ± 0.02	0.88 ± 0.02	2.22 ± 0.01	1.026 ± 0.02	1.082 ± 0.02	111.02 ± 0.02

and raise the content of active organic carbon. The reason is that due to the high-temperature hydrothermal hydrolysis of food waste, SHF is rich in a large amount of small molecular available carbon, such as amino acids and glucose. Its application to the soil has a significant promoting effect on improving the activity of soil organic carbon. The long-term field experiments by Xu et al. (2006) and the organic fertilizer experiment by Li et al. (2013) both showed that the application of organic fertilizer or organic materials combined with chemical fertilizers helps to increase the content of active organic carbon in different forms in the soil and correspondingly improves the soil carbon pool management index. Other research has shown (Zhang et al. 2012) that applying only chemical fertilizers is generally not conducive to the improvement of the soil carbon pool management index. However, the soil organic carbon content in treatments with chemical fertilizer application is generally higher than in the no-fertilizer control, indicating that the application of chemical fertilizers has a certain effect on maintaining soil organic carbon but is not conducive to the management of the soil carbon pool, which is consistent with the results of this study.

4.1.7 Summary

This section takes traditional organic fertilizer (TF) and chemical fertilizer (HF) as reference objects. It elucidates the influence mechanism of the synergistic and high-efficiency soil conditioner on humus formation in impoverished soil and evaluates its effect on the carbon pool quality management of impoverished cultivated land. The investigation focuses on the composition of soil aggregate structure and nutrient distribution, the changes in precursor components during humus formation, the evolution of functional groups, and the rate of soil humus formation. The main research conclusions are as follows:

(1) The application of the synergistic and high-efficiency soil conditioner increased the organic matter of impoverished soil by 3.44 times and 4.43 times compared to the application of organic fertilizer and chemical fertilizer, respectively. This indicates that the ability of the synergistic and high-efficiency soil conditioner to enhance the organic carbon pool storage in impoverished soil is significantly higher than that of traditional fertilization methods.
(2) Under the dual synergistic regulation of macromolecular humic carbon and small-molecule readily available carbon, the proportion of 0.05–1 mm mechanically stable aggregates in the soil increased, which is beneficial for the soil's retention of nutrient elements. It also increased the total organic carbon content of 0.25–2 mm size-class aggregates, contributing to soil carbon sequestration and promoting the enhancement of aggregate stability.
(3) Compared with the unfertilized treatment group, the amino acid content was highest in HF-S, followed by SHF-S. However, within the application period, the rate of amino acid reduction was highest in HF-S and lowest in SHF-S. The synergistic and high-efficiency soil conditioner can effectively improve the soil's retention capacity for amino acid nitrogen. The application of organic fertilizer and the synergistic and high-efficiency soil conditioner increased soil polysaccharide content by 27.78% and 36.62%, respectively, while the application of chemical fertilizer decreased it by 0.22 mg g^{-1}. The effect on reducing sugar content was consistent with that on polysaccharides, indicating that the synergistic and high-efficiency soil conditioner has a potential growth-promoting ability for the microbial biomass that cements soil particles into water-stable aggregates.
(4) Overall, the degree of DOM humification in TF-S was higher than in SHF-S and HF-S. However, the volume integral growth rate of fluorescent region V in SHF-S was 3.52% higher than in TF-S, indicating that the synergistic and high-efficiency soil conditioner has a stronger promoting rate for soil humification, which is conducive to the formation of soil humus. The degree of aliphatization

of soil humic acid was in the order of SHF-S > TF-S > HF-S. The application of the synergistic and high-efficiency soil conditioner is beneficial for increasing the activity of soil humic acid.

(5) The carbon pool management index of SHF-S was 1.09 times and 1.36 times that of the TF-S and HF-S treatment groups, respectively. The application of the synergistic and high-efficiency soil conditioner is conducive to the accumulation of soil organic carbon and significantly increases the content of active organic carbon.

4.2 Long-Term Effects on the Dynamics of Active Carbon Pools and Microbial Community Succession in Different Tillage Layers

The rational application of exogenous nutrients can directly or indirectly affect the physicochemical properties of soil and increase the content of available nutrients, making it an effective measure for increasing the storage of the soil organic carbon pool. At the same time, the rational application of exogenous substances can improve the ecological service functions of the microbial community, optimize the circulation of nutrients and energy flow in the soil, and adjust the dynamic storage of the soil organic carbon pool. Therefore, the rational application of exogenous nutrients can become an effective means of enhancing soil fertility and quality by effectively regulating the organic carbon pool.

The dynamics and transformation of soil organic carbon is a slow and complex process, and the evolutionary trend of active organic carbon components in the organic carbon pool is an important indicator for studying carbon pool transformation. Under the application of exogenous nutrients, the fixation potential of the soil's active organic carbon pool and the evolutionary trend of carbon components are influenced by factors such as fertilization methods, cropping systems, and soil types. Traditional fertilization methods often aim for the maximization of crop yield in the short term, and inorganic fertilizers such as chemical fertilizers have always been the first choice for fertilizing cultivated land due to their high nutrient content and rapid fertilizing effect. However, some studies suggest that the sole application of inorganic fertilizers leads to a single type of nutrient being added to the soil and can even cause soil pollution, soil degradation, and reduced grain yields (Ju et al. 2009; Myint et al. 2009). Some scholars have found that the application of inorganic fertilizers can promote the mineralization rate of the soil's native organic carbon pool as well as newly added organic carbon, leading to a loss of soil organic carbon (Jiang et al. 2014). However, other studies have found that NPK compound fertilizers and even the sole application of nitrogen fertilizer can significantly increase soil organic carbon content (Khan et al. 2007). Many research findings indicate that the application of organic fertilizers can significantly increase the storage of the soil organic carbon pool, especially the active organic carbon pool (Zhang et al. 2016, 2009, 2014). However, the above research conclusions are mostly based on small-scale simulation experiments, and their results focus more on the cumulative effects under short-term fertilization, lacking an explanation of the evolution of the soil organic carbon pool under long-term fertilization. Long-term fertilization is a more common practice in farming. Therefore, it is necessary to explore the dynamic changes of the soil organic carbon pool and the evolution of carbon components under long-term fertilization. Compared with traditional fertilization methods, it is currently unclear what differences exist in the impact of the synergistic and high-efficiency soil conditioner, as a new type of exogenous substance, on the soil's active organic carbon pool.

Effective application methods of exogenous nutrients can increase soil fertility by increasing the content of nutrients available to crops, thereby affecting the diversity, species abundance, and community structure composition of the microbial community. Moreover, after a long period of applying exogenous nutrients, the adjustment of the soil microbial community structure tends to be balanced and optimized. Therefore, the successional changes in the microbial community can serve as an important indicator reflecting changes in soil fertility. It has been reported that the long-term application of chemical fertilizers reduces soil biomass and enzyme activity (Ramirez et al. 2012). The application of inorganic nitrogen fertilizer can significantly change the structural composition of the soil microbial community but has little effect on the abundance of the microbial community (Paungfool-lonhienne et al. 2015). In contrast, reports related to organic fertilizers show that they have positive effects on soil microbial biomass, enzyme activity, and microbial functional diversity (Graham and Haynes 2005). Although the influence and regulatory mechanisms of traditional fertilization on microbial communities are relatively clear, it is still unknown how the synergistic and high-efficiency soil conditioner in this study, as a special exogenous nutrient, regulates and affects soil microorganisms. Furthermore, how the changes in the microbial community are coupled with the evolution of the organic carbon pool still requires further research.

Impoverished cultivated land is constrained by factors such as a shallow topsoil layer, poor soil structure, low organic matter content, and unbalanced nutrients, which restrict crop growth. Therefore, in this section, we conducted a long-term (3-year) field experiment, comparing the synergistic and high-efficiency soil conditioner with the inorganic chemical fertilizers commonly used in crop fertilization. By analyzing changes in indicators such as soil physicochemical properties, soil organic carbon components, enzyme activities, microbial communities, and the abundance of key microorganisms, we aim to reveal the long-term regulatory effects of the synergistic and high-efficiency soil conditioner on the dynamics of active carbon pools and microbial community succession in different soil layers of impoverished soil.

4.2.1 Changes in Soil Organic Matter and Nutrients in Different Tillage Layers

4.2.1.1 Soil Organic Matter Content

The field experiment was conducted at the demonstration base of the Chinese Research Academy of Environmental Sciences. The area has an annual average temperature of 16.3 °C, an annual average rainfall of 1196.8 mm, and an annual average relative humidity of 85%. The prevailing winds are northeasterly and southwesterly. The preparation method for the tested synergistic and high-efficiency conditioner (SHF) is described in Chap.3 of this book. The tested chemical fertilizer (HF) was a commercial NPK ternary compound fertilizer, with total nitrogen, phosphorus pentoxide, and potassium oxide each accounting for 15%. The soil type at the experimental site was yellow soil, and its physicochemical properties are shown in Table 4.7. The land underwent two years of soil homogenization treatment before the experiment began. The test crop was citrus (Shiranui mandarin, hereinafter referred to as "Shiranui"), cultivated using traditional methods. Fertilization, water management, and other measures followed local traditional planting practices.

The experiment included four treatment groups: original soil (CK), a group with no external nutrient application for 3 years (S3Y), a group with inorganic fertilizer application for 3 years (CF3Y), and a group with the synergistic and high-efficiency conditioner application for 3 years (SHF3Y). Each treatment was replicated three times, for a total of 12 plots, with each plot measuring 40 m^2 (5 m × 8 m). The application rate of SHF was 420 kg ha^{-1}, and the application rate of CF was 380 kg ha^{-1}, calculated based on the total nitrogen content as the standard. The test crop, citrus, was cultivated using traditional methods. The timing of fertilization, irrigation management, and other measures followed local traditional planting practices. Except for the absence of external nutrient application, the other land management conditions for the S3Y group were identical to those for the CF3Y and SHF3Y treatment groups.

As shown in Table 4.8, overall, compared to S3Y, the application of CF and SHF for 3 years increased the soil organic matter content in the tillage layer. The application of CF and SHF caused a significant "surface accumulation" of soil organic

Table 4.7 Basic properties of the test soil B

Physicochemical property (unit)	Value
pH	7.35 ± 0.13
Total nitrogen (g kg^{-1})	0.81 ± 0.24
Alkali-hydrolyzable nitrogen (mg kg^{-1})	11.58 ± 1.03
Total phosphorus (g kg^{-1})	0.41 ± 0.02
Available phosphorus (mg kg^{-1})	13.27 ± 4.22
Total potassium (g kg^{-1})	1.66 ± 0.77
Available potassium (mg kg^{-1})	87.31 ± 5.07
Organic matter (g kg^{-1})	8.79 ± 0.45

matter, meaning it primarily accumulated in the 0–20 cm tillage layer. In the 0–20 cm tillage layer, the soil organic matter content in the CF3Y and SHF3Y groups was 2.7 g kg^{-1} and 3.18 g kg^{-1} higher than in the S3Y group, respectively. In contrast, the organic matter content in the 0–20 cm tillage layer of the control group S3Y was 2.43 g kg^{-1} lower than that of CK. This indicates that under no-tillage and no-application conditions, soil organic matter is lost through mineralization, and the effect of SHF on the accumulation of organic matter in the 0–20 cm tillage layer is greater than that of CF. In the 20–30 cm and 30–40 cm soil layers, compared to CK, the soil organic matter content in the CF3Y group was slightly lower than CK, whereas the SHF3Y group was significantly higher than both S3Y and CK. This demonstrates that the application of the SHF conditioner can effectively fix carbon content in the sub-tillage and plow layers of the soil. This is because SHF is rich in small-molecule organic carbon, which can promote the formation of soil aggregates by regulating the composition of complexes and micro-aggregates, thereby significantly increasing the organic matter content in the soil tillage layer. While the application of CF had some effect on the accumulation of soil organic matter, it failed to restore it to the initial level.

4.2.1.2 Soil Nutrient Content

Compared to CK, the nutrient content in the S3Y treatment group decreased, indicating that nutrient loss occurs in barren soil without external supplementation. The application of CF and SHF to the soil caused a slight increase and decrease in soil pH, respectively. Additionally, both significantly increased the soil's total nitrogen, alkali-hydrolyzable nitrogen, total phosphorus, available phosphorus, total potassium, and available potassium content. The average total nitrogen content in the SHF3Y treatment group was 58.86% and 11.06% higher than the control group (S3Y) and the inorganic fertilizer group (CF3Y), respectively. The alkali-hydrolyzable nitrogen content was 99.41% and 2.7% higher than these two treatment groups, respectively (Table 4.8). This indicates that SHF can significantly increase the soil's nitrogen fixation capacity. This is mainly because SHF is rich in small-molecule substances, with most nutrients existing in an organic state. Through physicochemical processes and microbial fermentation and decomposition in the soil, these nutrients are gradually released and have high mobility in the soil profile. Therefore, SHF application still has a significant ameliorating effect on barren soil in deeper tillage layers. Research has shown that there is a significant positive correlation between the content of total nitrogen and alkali-hydrolyzable nitrogen and soil organic matter, and they follow the same trend, which is consistent with the findings of this study (Li et al. 2015). The total phosphorus content in the inorganic fertilizer group was not significantly different from the SHF conditioner group. However, the total potassium and available potassium content were higher than in the SHF conditioner group, which is consistent with the research findings of Chen et al. (2014).

The results for nutrient content in different soil layers showed that, compared to the application of inorganic fertilizer, the SHF conditioner effectively increased the

Table 4.8 Physicochemical characteristics of different soil profile layers

Treatment group	Tillage layer (cm)	pH	Total nitrogen (g kg^{-1})	Alkali-hydrolyzable nitrogen (mg kg^{-1})	Total phosphorus (g kg^{-1})	Available phosphorus (mg kg^{-1})	Total potassium (g kg^{-1})	Available potassium (mg kg^{-1})	Organic matter (g kg^{-1})
CK	0–20	7.35 ± 0.03	0.81 ± 0.24	11.58 ± 1.03	0.41 ± 0.02	13.27 ± 4.02	1.66 ± 0.77	87.31 ± 5.07	8.79 ± 0.45
	20–30	7.38 ± 0.03	0.78 ± 0.03	10.16 ± 1.05	0.37 ± 0.06	12.93 ± 2.52	1.24 ± 0.03	81.12 ± 6.44	7.44 ± 0.14
	30–40	7.43 ± 0.02	0.70 ± 0.05	9.26 ± 0.55	0.33 ± 0.07	12.04 ± 1.04	1.17 ± 0.28	80.29 ± 5.33	3.38 ± 0.84
S3Y	0–20	6.82 ± 0.07	0.70 ± 0.05	8.94 ± 0.22	0.35 ± 0.05	10.81 ± 1.28	1.27 ± 0.11	77.17 ± 6.06	6.36 ± 1.01
	20–30	6.83 ± 0.03	0.53 ± 0.02	8.49 ± 0.24	0.35 ± 0.02	9.95 ± 0.24	1.01 ± 0.02	74.36 ± 4.74	7.12 ± 0.09
	30–40	6.94 ± 0.02	0.35 ± 0.04	8.08 ± 0.44	0.30 ± 0.03	10.06 ± 1.04	0.96 ± 0.13	74.31 ± 2.33	3.19 ± 0.10
CF3Y	0–20	7.41 ± 0.02	0.82 ± 0.13	16.59 ± 1.23	0.49 ± 0.10	17.30 ± 0.53	1.62 ± 0.85	86.51 ± 3.66	9.06 ± 1.02
	20–30	7.7 ± 0.05	0.78 ± 0.02	16.43 ± 1.12	0.47 ± 0.05	16.59 ± 2.01	1.50 ± 0.02	86.22 ± 5.05	7.21 ± 1.06
	30–40	7.81 ± 0.02	0.66 ± 0.11	16.50 ± 0.25	0.46 ± 0.05	14.11 ± 1.77	1.51 ± 0.07	86.63 ± 3.21	3.14 ± 0.04
SHF3Y	0–20	7.25 ± 0.06	0.87 ± 0.04	17.15 ± 0.53	0.48 ± 0.07	17.44 ± 1.02	1.54 ± 0.06	83.36 ± 3.58	9.54 ± 0.39
	20–30	7.26 ± 0.04	0.83 ± 0.07	16.92 ± 1.02	0.45 ± 0.09	17.14 ± 1.12	1.41 ± 0.09	80.18 ± 2.44	7.89 ± 1.11
	30–40	7.36 ± 0.03	0.81 ± 0.07	16.80 ± 1.75	0.46 ± 0.09	16.88 ± 2.09	1.27 ± 0.02	75.25 ± 3.29	3.51 ± 0.38

content of total nitrogen, alkali-hydrolyzable nitrogen, and available phosphorus in the sub-tillage layer (20–30 cm) and the plow layer (30–40 cm). This indicates that for non-surface soil, the SHF conditioner still has good nitrogen and phosphorus retention effects.

4.2.1.3 Changes in Organic Carbon Fractions

Soil organic carbon fractions are significantly affected by the input of external nutrients. Among these, soil active organic carbon has the highest bioavailability and is easily utilized directly by soil microorganisms and plants. It plays an important role in maintaining soil fertility and balancing the fluctuations of the soil organic carbon pool.

In Fig. 4.24, within the 0–20 cm tillage layer, the Dissolved Organic Carbon (DOC), Microbial Biomass Carbon (MBC), and Labile Organic Carbon (LOC) contents in the SHF3Y treatment group were 227.74 mg kg^{-1}, 277.11 mg kg^{-1}, and 3.68 g kg^{-1}, respectively, which were significantly higher than their contents in the 20–30 cm and 30–40 cm tillage layers. This indicates that the distribution of soil organic carbon exhibits a relatively clear phenomenon of "surface accumulation". The CF3Y soil profile showed the largest variation in organic carbon fractions. Compared to the 0–10 cm tillage layer, the DOC, MBC, LOC, and Particulate Organic Carbon (POC) contents in the 30–40 cm tillage layer decreased by 63.33%, 81.61%, 66.67%, and 61.16%, respectively.

The average DOC content distribution from highest to lowest was: SHF3Y (169.67 mg kg^{-1}) > S3Y (135.35 mg kg^{-1}) > CK (114.17 mg kg^{-1}) > CF3Y (107.65 mg kg^{-1}), with significant overall differences ($P < 0.05$). The DOC content in the SHF3Y and CF3Y groups varied greatly with profile depth. In the 30–40 cm tillage layer, the DOC contents of SHF3Y and CF3Y were 0.37 and 0.46 times that of the 0–20 cm tillage layer, respectively, and the DOC in SHF3Y was significantly higher than in CF3Y ($P < 0.05$). This suggests that the hydrothermal conditions in the topsoil are better, leading to active microbial metabolism. Additionally, crop litter contributes to the carbon source reserves in the topsoil, making the DOC in the surface layer significantly higher than in the lower layers. It also indicates that SHF is more conducive to the transformation and circulation of inorganic substances in the soil and has a potential promoting effect on the growth and reproduction of soil microorganisms (Wu et al. 2013).

MBC actively participates in the soil nutrient cycle, can be directly absorbed by crops as a nutrient source, and is an important indicator for evaluating the biological properties of soil quality. The MBC content in the SHF3Y group was significantly higher than in the other treatment groups ($P < 0.05$). In the 0–20 cm tillage layer, the MBC content distribution from highest to lowest was: SHF3Y (277.11 mg kg^{-1}) > CF3Y (183.54 mg kg^{-1}) > CK (112.14 mg kg^{-1}) > S3Y (71.26 mg kg^{-1}). In the 20–30 cm tillage layer, the MBC content of SHF3Y was 5.9 times that of S3Y, indicating that the application of SHF significantly increased microbial activity in the 20–30 cm soil layer. In the 30–40 cm tillage layer, the MBC content of SHF3Y was 3.72, 7.04,

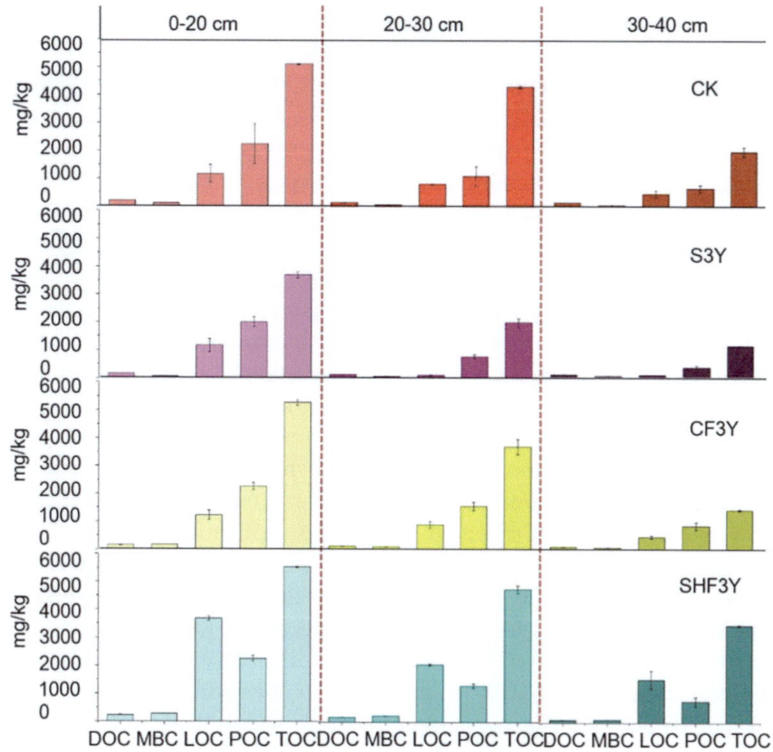

Fig. 4.24 Dynamic changes of soil organic carbon fractions in different layers

and 3.44 times that of CK, S3Y, and CF3Y, respectively. This demonstrates that the application of SHF accelerates the rate at which microorganisms decompose organic carbon sources and enhances assimilation. Compared to CF, SHF can increase soil nutrients, ensure a higher microbial biomass, and have the potential function of improving the soil microbial community structure (Ingram et al. 2005).

As an important component of soil organic carbon, LOC can reflect minor changes in soil carbon content and is directly involved in soil biochemical transformation processes. It can better reflect changes in soil quality, nutrient cycling, and transformation rates. As an energy source for the life activities of soil microorganisms, it plays a crucial role in driving the distribution of soil nutrients. The LOC content in all treatment groups varied significantly with profile depth ($P < 0.05$). The contribution rates of LOC content to Total Organic Carbon (TOC) in the CF3Y group were 23.39% (0–20 cm), 23.69% (20–30 cm), and 29.76% (30–40 cm), respectively. In the SHF3Y group, the contribution rates of LOC content to TOC in each tillage layer were 2.81, 1.82, and 1.49 times those of CF3Y, respectively. This is consistent with the trend of soil organic matter, mainly because changes in the soil carbon pool capacity primarily occur in the soil LOC carbon pool. Therefore, LOC can indicate early changes in soil organic matter. In the 0–20 cm tillage layer, the LOC contents in the CF3Y and SHF3Y groups increased by 5.12% and 214.5% compared to CK, respectively. SHF significantly enhanced the surface accumulation of LOC in barren farmland soil. In the 30–40 cm tillage layer, the LOC content in CF3Y decreased by 8.89% relative to CK, while in SHF3Y it increased by 244.4%. The short-term fluctuations and changes in soil organic carbon mainly occur in the easily oxidizable and decomposable parts, indicating that the application of the synergistic and high-efficiency conditioner has a significantly higher ability to promote the short-term turnover rate of soil organic carbon than chemical fertilizer (Zou et al. 2005).

The main sources of POC include undecomposed residues from the above-ground parts and roots of soil plants. The composition and content of POC in the soil can indirectly reflect the soil's carbon sequestration capacity and changes in soil quality. In the 0–20 cm tillage layer, the distribution of Particulate Organic Carbon (POC) content from highest to lowest was: CF3Y (2.26 g kg^{-1}) > SHF3Y (2.25 g kg^{-1}) > CK (2.24 g kg^{-1}) > S3Y (2.01 g kg^{-1}). Although no-tillage can promote soil POC accumulation under certain conditions, it generally has a lag effect, with the increase in POC becoming apparent only after 5 to 10 years; therefore, the POC content in the S3Y group was lower. Compared to CK, the application of CF and SHF for 3 years had little effect on the POC content of the barren soil. In the 20–30 cm tillage layer, compared to CK, the POC contents in the CF3Y and SHF3Y groups were higher by 0.42 g kg^{-1} and 0.19 g kg^{-1}, respectively, with contribution rates to the Total Organic Carbon (TOC) content of 41.66% and 27.25%, respectively. This indicates that, compared to SHF, the application of chemical fertilizer increased the proportion of sequestered organic carbon fractions, which tends to stabilize the soil organic carbon. At the 30–40 cm depth, the POC contents in the CK, S3Y, CF3Y, and SHF3Y treatment groups were lower than the surface layer by 1.6 g kg^{-1}, 1.68 g kg^{-1}, 1.45 g kg^{-1}, and 1.48 g kg^{-1}, respectively.

The soil TOC content in all treatment groups showed a decreasing trend with increasing profile depth. This is mainly because as the tillage layer deepens, the input of organic matter, such as crop roots and root exudates, decreases, which in turn reduces the circulation and transformation of soil organic carbon. In the CK and S3Y treatments, TOC decreased significantly with depth in the tillage layer. This shows that the frequent turning of the tillage layer under traditional cultivation destroys soil aggregates. In the absence of external organic matter input, the easily mineralizable organic carbon in the soil is continuously lost through mineralization, and this loss is compounded by increased soil respiration caused by human tilling disturbance, leading to a decline in soil organic carbon content. In the 30–40 cm tillage layer, the TOC contents for CK, S3Y, and CF3Y were 1.965 g kg^{-1}, 1.10 g kg^{-1}, and 1.38 g kg^{-1}, respectively. This suggests that the application of CF has little effect on organic carbon sequestration in the deep tillage layer of barren soil. In contrast, the average TOC content of SHF3Y was 1.78, 3.17, and 2.53 times that of CK, S3Y, and CF3Y, respectively, which demonstrates that the application of the synergistic and high-efficiency conditioner has a significant promotional effect on organic carbon sequestration in barren soil.

4.2.2 Soil Microbial Community Succession

4.2.2.1 Bacterial Community Diversity

As shown in Fig. 4.25, the coverage index for all samples was around 96%, indicating that the sequencing depth of bacterial genes in the soil for each treatment in this study was very sufficient. The Chao1 index and the rarefaction curve (Observed OTUs) index were highest for SHF3Y and lowest for CF3Y. This indicates that the application of the synergistic and high-efficiency conditioner for 3 years increased the abundance of the soil bacterial community, whereas the application of chemical fertilizer decreased it. Besides, the obtained rarefaction curves reflect the sampling depth of the samples and can be used to evaluate whether the sequencing volume could cover all taxa. The results of the Shannon diversity index were consistent with the rarefaction curves, indicating that the synergistic and high-efficiency conditioner can increase bacterial diversity in barren soil, while the application of chemical fertilizer reduces it.

4.2.2.2 Changes in Soil Bacterial Community Composition

A total of 242,400 sequences and 4,080 OTUs were detected across the four treatment groups, as shown in Fig. 4.26. Among these, 1,976 OTUs were common to CK, S3Y, CF3Y, and SHF3Y. These OTUs, as dominant species, accounted for 92.39% of the total species abundance. The CK treatment had 103 unique species, accounting for 0.13% of the total species abundance, belonging to *Armatimonadetes*, *Deinococcus-Thermus*, *Proteobacteria*, *Acidobacteria*, *Bacteroidetes*, *Cyanobacteria*, *Planctomycetes*, and *Firmicutes*. The S3Y treatment had 84 unique species, accounting for 0.09% of the total species abundance. *Actinobacteria*, *Chloroflexi*, *Gemmatimonadetes*, and *Latescibacteria*, as newly emerged dominant species, replaced *Armatimonadetes*, *Deinococcus-Thermus*, and *Cyanobacteria*. This suggests that as nutrients were consumed, eutrophic microorganisms decreased, while oligotrophic microorganisms such as *Chloroflexi* began to occupy the ecological niche.

The highest number of unique species was detected in the CF3Y treatment group, with a total of 148 species belonging to 24 bacterial phyla, accounting for 0.20% of the total abundance, which was significantly higher than the CK and S3Y groups. Among them, OTU1517, OTU1994, OTU1192, and OTU1891 had the highest relative abundance, belonging to Planctomycetia (OTU1517, OTU1192), *Betaproteobacteria* (OTU1891), and *Acidobacteria* (OTU1994), respectively. Among these, *Planctomycetia* (OTU1517, OTU1192) had the highest relative abundance. It has been reported that the currently discovered anaerobic ammonia-oxidizing bacteria all belong to *Planctomycetia* and are involved in the soil's anaerobic ammonia oxidation process, which can convert effective soil nitrogen into nitrogen gas for release. In this study, the application of chemical fertilizer may have promoted soil nitrogen mineralization, which is unfavorable for soil nitrogen fixation and crop growth.

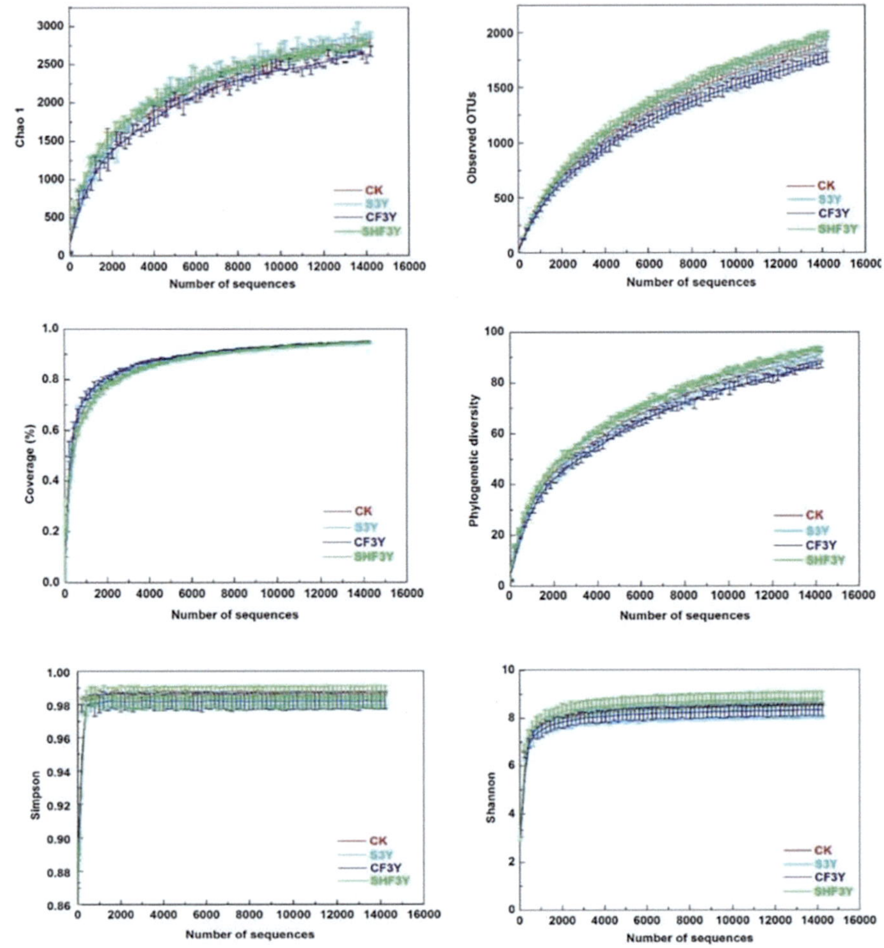

Fig. 4.25 Diversity indices of the same amount of sequences from original soil (CK), soil without fertilizers/soil conditioners for 3 years (S3Y), soil with chemical fertilizer for 3 years (CF3Y) and soil with SHF soil conditioner for 3 years (SHF3Y). Data for all samples were based on three soil layers (0–20 cm, 20–30 cm and 30–40 cm)

SHF3Y had 82 unique species, accounting for 0.16% of the total species abundance. Among them, OTU853, OTU467, OTU1732, and OTU1456 were dominant in the entire clone library with the highest relative abundance, belonging to *Bacilli* (OTU467) and *Ktedonobacteria* (OTU853, OTU1732, OTU1456), respectively. Among these, *Ktedonobacteria* (OTU853, OTU1732, OTU1456) had the highest relative abundance. Certain genera and species of *Ktedonobacteria* have strong cellulose degradation capabilities, indicating that the application of the synergistic and high-efficiency conditioner may play a role in promoting the decomposition of plant and animal residues in the soil.

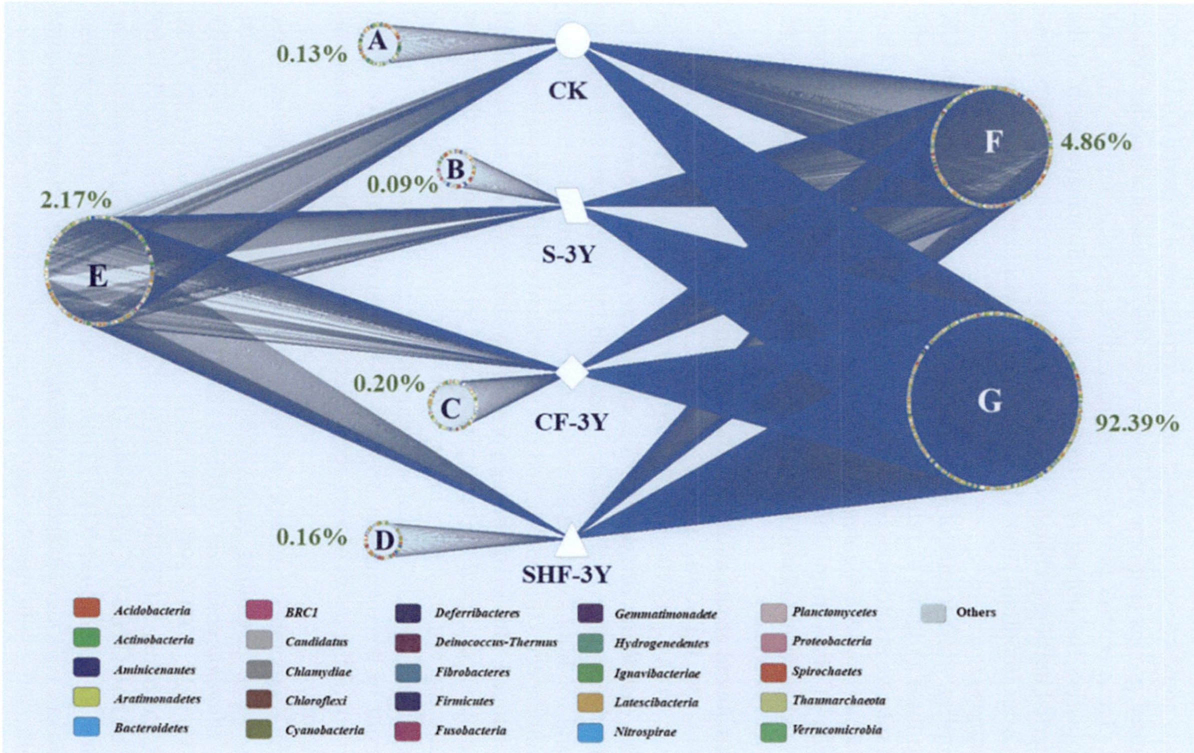

Fig. 4.26 An OTU network map showing the interactions of the OTUs among different treatments, at the phylum level. CK, S3Y, CF3Y, SHF3Y represented original soil (CK), soil without fertilizer/soil conditioners for 3 years (S3Y), soil with chemical fertilizer for 3 years (CF3Y) and soil with SHF soil conditioner for 3 years (SHF3Y). Each point represents one independent bacterial phylum. Phyla in the left column were shared with two soil samples; those in the middle were unique to one soil sample; phyla in the upper right were shared with three soil samples; whereas those in the lower right belonged to all samples

4.2.2.3 Analysis of Indicator Species Succession

The screening of indicator species was calculated using the Dufrene-Legendre indicator species analysis method, invented by Dufrene and Legendre in 1997. This method identifies indicator species based on the correlation of bacterial OTUs with different treatment groups (Dufrene and Legendre 1997). Indicator species reflect environmental characteristics through inter-group differences in sample units and can be used to assess the impact of external environmental changes (such as the addition of external nutrients, changes in tillage methods, etc.) on soil microorganisms, serving as an important means of monitoring environmental and biological community changes.

As shown in Fig. 4.27, a total of 27 indicator species were detected in the CK group. The top three in terms of relative abundance were OTU564, OTU633, and OTU513, with relative abundances of 4.95%, 3.47%, and 3.14%, respectively. They belong to *Firmicutes* and *Sphingobacteriales*, respectively. It has been reported that *Sphingobacteriia sp.* plays an important role in the degradation of polycyclic aromatic hydrocarbons such as benz[a]acridine, pentacene, and acenaphthylene (Zhao et al. 2011).

In the S3Y treatment group, a total of 6 indicator species were detected (OTU973, OTU1734, OTU1836, OTU3402, OTU2383, OTU2773), with relative abundances of 1.32%, 0.99%, 0.83%, 0.67%, 0.67%, and 0.67%, respectively. They belong to *Planctomycetes*, *Candidate division WPS-2*, and *Bacteroidetes*, respectively. The significant reduction in microbial indicator species in the unfertilized soil suggests that under natural conditions without the addition of external nutrients, the ecological niches of indicator species that had an "inertial" dependence on the original nutrients (food source) changed. This led to the high probability that previously non-dominant microbial populations occupied the ecological niches left after the extinction of the original indicator species, showing a higher correlation with barren soil. Therefore, the reason for the decrease in indicator species in the natural state of barren soil may be the result of the interaction between microorganisms and environmental conditions, as well as the result of long-term selection by environmental conditions on microorganisms.

In the CF3Y treatment group, a total of 36 indicator species were detected (Fig. 4.27). Excluding unclassified OTUs, the other indicator species belonged to *Planctomycetaceae* (OTU1941, OTU3394, OTU710, OTU893, OTU2362, OTU1524, OTU2291, OTU1517), *Alphaproteobacteria* (OTU1548, OTU2589), Chloroflexi (OTU3685, OTU2561), *Acidobacteria* (OTU1469, OTU1545, OTU1124, OTU3306, OTU1446, OTU2548, OTU1994), *Gemmatimonas* (OTU2755), and *Actinobacteria* (OTU3278). *Planctomycetaceae* had the highest abundance ratio in CF3Y. Certain species of *Planctomycetaceae* can directly oxidize ammonia to nitrogen gas in the absence of oxygen. Therefore, the application of chemical fertilizer promoted the proliferation of soil *Planctomycetes*, allowing them to occupy a dominant ecological niche and accelerating the mineralizational loss of soil nitrogen.

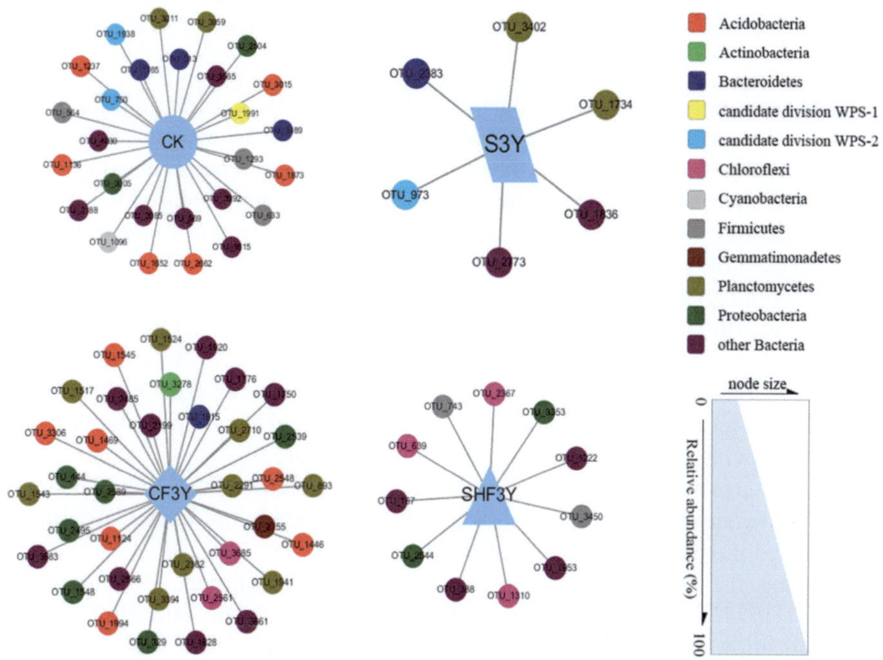

Fig. 4.27 Indicator species by treatment regime. Each small circle connected to the line represents an OTU, and the color shown in each of the OTU's circles indicates the phylum to which the OUT belonged. CK, S3Y, CF3Y, SHF3Y represented original soil (CK), soil without fertilizers/soil conditioners for 3 years (S3Y), soil with chemical fertilizer for 3 years (CF3Y) and soil with SHF soil conditioner for 3 years (SHF3Y)

In the SHF3Y treatment group, a total of 11 indicator species were detected. Among them, OTU167, OTU388, OTU1222, and OTU1953 were unclassified bacteria. The remaining species belonged to *Gammaproteobacteria* (OTU3353), *Ktedonobacter* (OTU639, OTU2367), *Sphaerobacter* (OTU1310), *Alicyclobacillus* (OTU743, OTU3450), and *Rhodospirillales* (OTU2544), respectively. Several scholars have studied and found that many genera in the *Enterobacteriaceae* family of *Gammaproteobacteria* have nitrogen-fixing functions, such as *Enterobacter cloaceae*, *Klebsiella planticola*, and *Klebsiella pneumoniae*. This indicates that the application of the synergistic and high-efficiency conditioner has great potential for promoting soil nitrogen fixation.

4.2.3 Summary

This chapter revealed the long-term effects of a synergistic and high-efficiency conditioner on the active organic carbon fractions, enzyme activities, microbial community structure changes, and indicator species succession in barren soil. The main research conclusions are as follows:

(1) Under natural conditions without the application of external nutrients, the degree of soil organic matter mineralization and loss was an average carbon loss of 12.52% from the surface layer to the plow layer. The application of the synergistic and high-efficiency conditioner for 3 years increased the average soil organic matter by 23.6%. The soil's effective nitrogen fixation capacity increased by 99%.
(2) The application of the synergistic and high-efficiency conditioner significantly increased the active organic carbon (with the exception of POC) and total organic carbon content in the barren soil. Compared to chemical fertilizer and no fertilization, the synergistic and high-efficiency conditioner can effectively promote the short-term turnover rate of soil organic carbon and has the potential function of improving the soil microbial community structure.
(3) The synergistic and high-efficiency conditioner increased the bacterial community diversity in the barren soil, whereas the application of chemical fertilizer reduced the bacterial diversity of the barren soil.
(4) Among the unique species in the CF3Y treatment group, *Planctomycetia* (OTU1517, OTU1192) had the highest relative abundance, indicating that the application of chemical fertilizer may have intensified the soil's anaerobic ammonia oxidation process, which is unfavorable for soil nitrogen fixation and crop growth. Among the unique species in the SFH3Y group, *Ktedonobacter* (OTU853, OTU1732, OTU1456) had the highest relative abundance. The application of the synergistic and high-efficiency conditioner may promote the decomposition of plant and animal residues in the soil by increasing the abundance of Ktedonobacter.
(5) The application of chemical fertilizer made *Planctomycetaceae* the most abundant indicator species in the CF3Y group; therefore, the chemical fertilizer promoted the proliferation of the soil Planctomycetes population, allowing it to occupy a dominant ecological niche and promoting the mineralization and loss of soil nitrogen. The indicator species for the SHF3Y treatment group were *Gammaproteobacteria* (OTU3353), *Ktedonobacter* (OTU639, OTU2367), *Sphaerobacter* (OTU1310), *Alicyclobacillus* (OTU743, OTU3450), and *Rhodospirillales* (OTU2544). The application of the synergistic and high-efficiency conditioner promotes soil cellulose degradation and soil nitrogen fixation by changing the successional dynamics of the aforementioned indicator species.
(6) Appropriately adding phosphorus and nitrogen sources or increasing the soil MBC and LOC content during the preparation of the synergistic and high-efficiency conditioner can promote the growth of indicator species in the SHF3Y group, which is beneficial for humus formation in barren soil.

4.3 Chapter Summary

By investigating the long-term effects of the synergistic and high-efficiency conditioner in different tillage layers of barren arable soil, this study proposed key controlling factors and environmental conditions for improving soil fertility, providing a reference basis for its engineering application. The main conclusions are as follows:

(1) The application of the synergistic and high-efficiency conditioner significantly increased the percentage content of 0.05–2 mm size aggregates, as well as the growth rate of organic carbon and total nitrogen. By promoting the formation of humus precursor substances including polysaccharides, reducing sugars, and carboxyl and phenolic hydroxyl compounds, it promoted soil humus formation, increasing the humification rate by 12.74%, which was significantly higher than the control group. The soil carbon pool management index for the soil treated with the synergistic and high-efficiency conditioner was 1.09 and 1.36 times that of the soils treated with organic fertilizer and chemical fertilizer, respectively.

(2) Under synergistic regulation, the abundance and diversity of the soil bacterial community increased. The synergistic and high-efficiency conditioner significantly increased the relative abundance of *Proteobacteria*, *Firmicutes*, and *Nitrospirae* in the soil. Correlation analysis showed that *Oxalobacteraceae* was greatly affected by traditional fertilizer, while the synergistic and high-efficiency conditioner could significantly increase the species abundance of *Lysobacter*. 50 days after the application of the conditioner, directionally amplifying the abundance of *Lysobacter* and *GP4* species is an effective way to increase the organic components of carboxyl, phenolic hydroxyl, polysaccharides, and reducing sugars in barren soil. When selecting traditional organic fertilizer as an external nutrient to condition barren soil, the application amount needs to be controlled so that the stimulation of *Microvirga* does not create feedback inhibition on the production of amino acids.

(3) In terms of long-term effects on barren arable soil, the application of the synergistic and high-efficiency conditioner significantly increased the soil organic matter and alkali-hydrolyzable nitrogen content, as well as the DOC, MBC, LOC, and TOC content in all soil tillage layers. The application of chemical fertilizer significantly increased the relative abundance of *Planctomycetaceae*, indicating that the application of chemical fertilizer has the potential to promote the mineralization and loss of nitrogen in barren arable soil. The synergistic and high-efficiency conditioner promoted an increase in the abundance of species related to nitrogen-fixing functions, such as *Gammaproteobacteria*, *Ktedonobacter*, *Sphaerobacter*, *Alicyclobacillus*, and *Rhodospirillales*.

References

Badia-Villas D, Girona-García A (2017) Soil humus changes with elevation in scots pine stands of the Moncayo Massif (NE Spain). Appl Soil Ecol 123:617–621

Bottinelli N, Angers DA, Hallaire V, Michot D, Guillou CL, Cluzeau D, Heddadj D, Menasseri-Aubry S (2017) Tillage and fertilization practices affect soil aggregate stability in a Humic Cambisol of Northwest France. Soil Tillage Res 170:14–17

Cambardella CA, Elliott ET (1993) Methods for physical separation and characterization of soil organic matter fractions. Geoderma 56(14):449–457

Chen H, Cao C, Kong L, Zhang C, Li W, Qiao Y, Du S, Zhao Z (2014) Study on wheat yield stability in huaibei lime concretion black soil area based on long-term fertilization experiment. Sci Agric Sin 47(13):2580–2590

Chepil WS (1951) Properties of soil which influence wind erosion: IV. State of dry aggregate structure. Soil Sci 72(5):387–402

Christensen BT (1992) Physical fractionation of soil and organic matter in primary particle size and density separates. Springer

Dou S (2010) Soil organic matter. Science Press

Dufrene M, Legendre P (1997) Species assemblages and indicator species: the need for a flexible asymmetrical approach. Ecol Monogr 67(3):345–366

Galantini J, Rosell R (2006) Long-term fertilization effects on soil organic matter quality and dynamics under different production systems in semiarid Pampean soils. Soil Tillage Res 87(1):72–79

Gigliotti G, Macchioni A, Zuccaccia C, Giusquiani PL, Businelli D (2003) A spectroscopic study of soil fulvic acid composition after six-year applications of urban waste compost. Agronomie 23(8):719–724

Graham MH, Haynes RJ (2005) Organic matter accumulation and fertilizer-induced acidification interact to affect soil microbial and enzyme activity on a long-term sugarcane management experiment. Biol Fertil Soils 41(4):249–256

Gupta VVSR, Germida JJ (1988) Distribution of microbial biomass and its activity in different soil aggregate size classes as affected by cultivation. Soil Biol Biochem 20(6):777–786

Ingram LJ, Schuman GE, Stahl PD, Spackman LK (2005) Microbial respiration and organic carbon indicate nutrient cycling recovery in reclaimed soils. Soil Sci Soc Am J 69(6):1737–1745

Jiang G, Xu M, He X, Zhang W, Huang S, Yang X, Liu H, Peng C, Shirato Y, Iizumi T (2014) Soil organic carbon sequestration in upland soils of northern China under variable fertilizer management and climate change scenarios. Glob Biogeochem Cycles 28(3):319–333

Ju XT, Xing GX, Chen XP, Zhang SL, Zhang LJ, Liu XJ, Cui ZL, Yin B, Christie P, Zhu ZL (2009) Reducing environmental risk by improving N management in intensive Chinese agricultural systems. Proc Natl Acad Sci USA 106(9):3041

Khan SA, Mulvaney RL, Ellsworth TR, Boast CW (2007) The myth of nitrogen fertilization for soil carbon sequestration. J Environ Qual 36(6):1821

Kostenko IV (2017) Relationships between parameters of the humus status of forest and meadow soils and their altitudinal position on the main Crimean range. Eurasian Soil Sci 50(5):515–525

Kristiansen SM, Schjønning P, Thomsen IK, Olesen JE, Kristensen K, Christensen BT (2006) Similarity of differently sized macro-aggregates in arable soils of different texture. Geoderma 137(1):147–154

Li C, Gao L, Li J, Liu B, Zou G, Sun Q, Xu J (2013) Impact of different organic fertilizers on different forms of organic carbon and carbon pool management index in a newly-built greenhouse vegetable soil. Acta Agriculturae Zhejiangensis 25(5):1079–1085

Li J, Xin J, Zhang H, Duan J, Ren Y, Sun N, Xu M (2015) Evolution characteristics of soil nutrients in the main rice production regions, the middle-lower reach of Yangtze River of China. J Plant Nutr Fertil 21(1):92–103

Lu J, Zheng F, Li G, Bian F, An J (2016) The effects of raindrop impact and runoff detachment on hillslope soil erosion and soil aggregate loss in the Mollisol region of Northeast China. Soil Tillage Res 161(161):79–85

Myint TZ, Sooksathan I, Kaveeta R, Juntakool S (2009) Effects of different organic amendments and chemical fertilizer on plant growth and grain yield of soybean on Pakchong soil series. Kasetsart J Nat Sci 43(3):432–441

Ramirez KS, Craine JM, Fierer N (2012) Consistent effects of nitrogen amendments on soil microbial communities and processes across biomes. Glob Change Biol 18(6):1918–1927

Rowe H, Withers PJA, Baas P, Chan NI, Doody D, Holiman J, Jacobs B, Li HG, Macdonald GK, Mcdowell R (2015) Integrating legacy soil phosphorus into sustainable nutrient management strategies for future food, bioenergy and water security. Nutr Cycl Agroecosyst 104(3):1–20

Wang J, Wang Y, Yao Y, Xu J, Fan J (2017) Effects of long-term fertilization on phosphorus retention and release of soil aggregates in upland red soils. Acta Pedol Sin 54(5):1240–1250

Wang W, Lai DYF, Wang C, Pan T, Zeng C (2015) Effects of rice straw incorporation on active soil organic carbon pools in a subtropical paddy field. Soil Tillage Res 152:8–16

Wang Z, Yang S, Zhao C, Bai J, Lou H, Chen K, Wu L, Dong G, Zhou Q (2016) Assessment of non-point source total phosphorus pollution from different land use and soil types in a mid-high latitude region of China. Water 2016(8):505

Wu J, Zhang J, Qian J, Huang J (2013) Intercropping grasses improve soil organic carbon content and microbial community functional diversities in Chinese hickory stands. Trans Chin Soc Agric Eng 29(20):111–117

Xu M, Yu R, Sun X, Liu H, Wang B, Li J (2006) Effects of long-term fertilization on labile organic matter and carbon management index (CMI) of the typical soils of China. Plant Nutr Fertil Sci 12(4):459–465

Zhang PP, Liu YJ, Xiao-Zhen PU, Zhang GJ, Jin W, Zhang WF (2016) Effects of residue management and fertilizer application mode on soil organic carbon pools in an oasis cotton region. Chin J Appl Ecol 27(11):3529–3538

Zhang QC, Shamsi IH, Xu DT, Wang GH, Lin XY, Jilani G, Hussain N, Chaudhry AN (2012) Chemical fertilizer and organic manure inputs in soil exhibit a vice versa pattern of microbial community structure. Appl Soil Ecol 57:1–8

Zhang W, Xu M, Wang B, Wang X (2009) Soil organic carbon, total nitrogen and grain yields under long-term fertilizations in the upland red soil of southern China. Nutr Cycl Agroecosyst 84(1):59–69

Zhao H, Dai S, Sun W, Wang G, Guo Y, Guo J (2011) Comparison of bacterial diversity of polluted and unpolluted sediment by brominated flame retardant. Acta Microbiol Sin 51(3):377–385

Zhao J, Chen S, Hu R, Li Y (2017) Aggregate stability and size distribution of red soils under different land uses integrally regulated by soil organic matter, and iron and aluminum oxides. Soil Tillage Res 167:73–79

Zhou H, Yu WT, Zhao Y (2014) Total and labile carbon in Alfisol soil amended with plant residual and livestock manure. Adv Mater Resh 988:411–415

Zou XM, Ruan HH, Fu Y, Yang XD, Sha LQ (2005) Estimating soil labile organic carbon and potential turnover rates using a sequential fumigation–incubation procedure. Soil Biol Biochem 37(10):1923–1928

Chapter 5
Amelioration of Secondary Soil Salinization Using a Synergistic and High-Efficiency Soil Conditioner

Abstract Greenhouse agriculture refers to crop cultivation conducted in an artificially protected environment, also known as protected cultivation. Currently, China has the largest area of agricultural facilities in the world, exceeding 85% of the world's total. Greenhouse cultivation is prone to problems such as soil compaction, nutrient imbalance, and secondary salinization (Yang 2018). In soils with secondary salinization, not only is nutrient utilization efficiency and microbial activity low, but crop yield and quality are also poor. Furthermore, after the aggregate structure of greenhouse soil is destroyed, it exacerbates soil compaction, affects gas–liquid circulation in the soil, and leads to the deterioration of the soil environment, severely hindering the green and sustainable development of greenhouse agriculture. At present, the use of bio-organic fertilizers or soil conditioners to improve soil quality has received widespread attention. There is an urgent need for efficient methods of co-applying organic fertilizers and soil conditioners to improve soil structure, reduce salt accumulation in the surface soil, balance soil nutrient content, improve the soil rhizosphere environment, and enhance soil quality. This chapter investigates the application of a soil conditioner prepared from food waste mixed with bio-organic fertilizer to secondarily salinized greenhouse agriculture soil. It studies the improvement effects of the combined application of organic fertilizer and soil conditioner on issues such as nutrient imbalance, high salt content, and low microbial community activity in greenhouse soils. The study elucidates the mechanism by which the combined fertilizer application ameliorates secondarily salinized soil, providing new ideas for its remediation.

5.1 Effects of the Synergistic and High-Efficiency Soil Conditioner on the Physicochemical Properties of Soil

The research site is located in Shouguang City, Shandong Province, China (36° 52′ N, 118° 54′ E), where the main local soil type is silty alluvial soil. The area has a typical temperate continental monsoon climate, with an average annual precipitation of 593.8 mm, an average annual temperature of 12.4 °C, and an average annual

J. Hou et al., *Food Waste Valorization: Preparation, Properties, and Application of the Synergistic Soil Conditioners*, https://doi.org/10.1007/978-981-95-5362-4_5

sunshine duration of 2548.8 h. The properties of the 0–20 cm soil layer before fertilizer treatment were as follows: soil organic matter (OM) content was 16.93 g kg^{-1}; total nitrogen (TN) content was 220 mg kg^{-1}; available phosphorus (AP) content was 64.49 mg kg^{-1}; available potassium (AK) content was 505 mg kg^{-1}; pH was 8.103; and total dissolved solids (TDS) content was 6.675 g kg^{-1}. The main soluble cations were K^+ and Ca^{2+}, and the main anions were SO_4^{2-} and NO_3^-. The soil was alkaline, and the TDS in the soil far exceeded 2 g kg^{-1}, classifying it as severely salinized soil.

The fertilizers used for soil amelioration included one type of bio-organic fertilizer (BOF) and two soil conditioners (Conditioner A and Conditioner B). The BOF was produced using edible fungus residue and animal protein powder as a substrate, with supplementary materials such as seaweed and plant ash added, followed by high-temperature fermentation. BOF contains a large amount of nutrients and trace elements required by crops. Conditioner A was produced using food residue and rice husks as raw materials, with the addition of a microbial inoculant, and underwent rapid high-temperature fermentation (70 °C, 18 h) in a bioreactor. The total organic matter in the resulting Conditioner A was about 80% (dry weight), of which about 60% was macromolecular humic carbon, and 20% was small molecular available carbon. Soil Conditioner B was mainly composed of minerals, plant hormones, and a microbial inoculant. The basic physicochemical properties of the different fertilizers are shown in Table 5.1.

Based on the properties of the soil and fertilizers, five treatment groups were designed: no fertilizer (CK), low-rate application of organic fertilizer alone (S-1), high-rate application of organic fertilizer alone (S-2), combined application of organic fertilizer and Conditioner A (M-1), and combined application of organic fertilizer and Conditioners A and B (M-2). The experimental design and fertilizer application rates are listed in Table 5.2. After uniform fertilizer application, the land was tilled with a rotary tiller, and the greenhouse was closed for 2–3 days before crop planting.

Simulation Experiment: To clarify the effect of the bio-organic fertilizer combined with the soil conditioners on the migration and transformation of salt ions in the soil, five sets of indoor simulation experiments were designed: no fertilizer (CK), organic fertilizer alone (T1), Conditioner A alone (T2), combined application of organic fertilizer and Conditioner A (T3), and combined application of organic fertilizer and Conditioners A and B (T4). The experimental design and fertilizer application rates

Table 5.1 Basic physicochemical properties of the different fertilizers

Fertilizer type	pH	TDS (g/kg)	TOC (g/kg)	TN (g/kg)	AP (mg/kg)	AK (mg/kg)
BOF	6.96 ± 0.22	19.87 ± 0.68	239.6 ± 8.15	22.21 ± 0.64	21.20 ± 0.51	732 ± 24.89
Conditioner A	6.78 ± 0.15	59.41 ± 2.14	340.6 ± 10.89	21.72 ± 0.72	88.34 ± 0.64	687 ± 22.67
Conditioner B	8.46 ± 0.28	18.95 ± 0.57	134.2 ± 4.43	4.06 ± 0.10	29.15 ± 0.76	1200 ± 45

Table 5.2 Fertilizer application rates for the different experimental groups in the field trial

Different fertilization methods	BOF (kg m^{-2})	Conditioner A (kg m^{-2})	Conditioner B (kg m^{-2})	TDS (g kg^{-1})
CK	–	–	–	0
S-1	1.19	–	–	315.33
S-2	4.79	–	–	1268.67
M-1	1.19	0.89	–	545.07
M-2	1.19	0.89	0.37	113.73

Note "–" indicates no fertilizer was applied

Table 5.3 Fertilizer application rates for the different experimental groups in the simulation experiment

Different fertilization methods	BOF (kg m^{-2})	Conditioner A (kg m^{-2})	Conditioner B (kg m^{-2})
CK	–	–	–
T1	2.45	–	–
T2	–	2.45	–
T3	1.23	1.23	–
T4	0.98	0.98	0.49

Note "–" indicates no fertilizer was applied

are listed in Table 5.3. The experiment was conducted in 2 L pots, each containing 1800 g of soil. The soil moisture content was maintained at 65% of the maximum field water holding capacity, and the pots were placed in a constant temperature incubator at 25 °C for a 60-day incubation period. Water was applied every 7 days, and the leachate was collected. Soil samples were taken at 1, 7, 14, 28, 40, and 60 days, collected from multiple points, mixed evenly, and air-dried for later analysis.

5.1.1 Changes in Soil pH, Bulk Density, and Electrical Conductivity Under Different Fertilization Regimes

In recent years, studies have found that the effectiveness of soil fertility is highest when the soil pH is between 6.5 and 7.5 (Chen et al. 2009). As can be seen from Table 5.4, the soil in all treatment groups was alkaline (pH between 7.5 and 8.5). Compared to CK, the pH value decreased in all fertilized treatments except for S-2. The decrease was most significant in S-1, which was 4.73% lower than CK. The pH values in the M-1 and M-2 groups were similar, decreasing by 2.87% and 1.51%, respectively, compared to CK. Qi et al. found that the long-term application of organic fertilizer significantly lowered soil pH, with the most significant effect on improving the fertility of saline-alkaline soils (Qi 2016). The research of Zhang et al. discovered that applying organic fertilizer can reduce soil pH and maintain the stability of the soil's acid–base buffering capacity, thereby keeping the soil pH within a range beneficial to soil health, which is consistent with the results of this study (Zhang et al. 2016). The application of organic fertilizer can lower soil pH, and its combined application with a soil conditioner is even more beneficial for reducing soil pH. This is partly due to the pH of the fertilizers themselves and partly because the decomposition of the organic fertilizer and conditioner can produce organic acids, leading to a decrease in soil pH. Additionally, the long-term application of animal manure compost can significantly increase soil pH. In this study, the excessive application of organic fertilizer led to an increase in soil pH, exacerbating soil salinization.

The effects of different fertilization methods on soil electrical conductivity are shown in Table 5.4. The electrical conductivity under the S-2 treatment (630.8 μS cm^{-1}) increased significantly, about 3 times that of CK (202.2 μS cm^{-1}). The changes in electrical conductivity for M-1 and M-2 were similar, at 249.2 μS cm^{-1} and 243.5 μS cm^{-1} respectively, representing increases of 23.24% and 20.43% compared to CK. Only in the S-1 treatment group did the electrical conductivity decrease.

The looser and more porous the soil, the smaller its bulk density; conversely, the more compact the soil, the larger its bulk density. As shown in Table 5.4, the bulk density in the fertilization treatments decreased compared to the control group, with the order of bulk densities being M-2 < M-1 < S-1 < S-2 < CK. The reduction was most significant in the M-1 and M-2 groups, with values of 1.16 g cm^{-3} and 1.09 g cm^{-3}, respectively, representing decreases of 29.27% and 33.54% compared to the control group. This indicates that applying organic fertilizer, especially when combined with a soil conditioner, helps to reduce bulk density, increase soil porosity, improve soil structure, enhance soil aeration, and create a better micro-ecological environment.

Table 5.4 Changes in soil pH, electrical conductivity, and bulk density under different fertilization methods

Different fertilization methods	pH	Electrical conductivity (μS/cm)	Bulk density (g/cm^3)
CK	8.02 ± 0.27 a	202.2 ± 9.301 c	0.56245 ± 0.016 a
S-1	7.641 ± 0.30 a	194.8 ± 7.597 c	0.56104 ± 0.020 a
S-2	8.043 ± 0.23 a	630.8 ± 23.339 a	0.541801 ± 0.022 a
M-1	7.798 ± 0.24 a	249.2 ± 12.25 b	0.55829 ± 0.022 a
M-2	7.899 ± 0.27 a	243.5 ± 7.062 b	0.522549 ± 0.026 a

Note Different lowercase letters indicate significant differences between treatments at the 5% probability level

5.1.2 Changes in Soil Total Nitrogen, Available Phosphorus, and Readily Available Potassium Under Different Fertilization Regimes

Regarding Total Nitrogen (TN), the content in all treatment groups was less than 500 mg kg^{-1}, which is considered an extremely deficient level. With the exception of S-2, all other fertilization treatments increased the TN content in the soil. The increase was most pronounced in the M-1 and M-2 groups, where the TN content was 1.59 and 1.41 times that of the control group, respectively (Fig. 5.1). The change in TN under the S-1 treatment was not significant, showing a 3.70% increase compared to the control group. This suggests that the combined application of bio-organic fertilizer and the soil conditioner is beneficial for nitrogen sequestration in the soil. On one hand, the organic fertilizer and conditioner contain abundant nutrients that can directly increase the nitrogen in the soil. On the other hand, Soil Conditioner A, which contains small-molecule substances, can activate microorganisms and promote plant root development, preserving nitrogen in the soil through microbial metabolism and plant transformation.

Overall, the Available Phosphorus (AP) and Available Potassium (AK) levels in all treatment groups were very high (> 40 mg kg^{-1} and > 200 mg kg^{-1}, respectively) (Fig. 5.2). Fertilization did not have a significant effect on soil AP and AK. The AP in the S-1 and M-1 groups increased by 16.98% and 93.59%, respectively, compared to CK. The AK in the S-1 and M-1 groups increased by 21.73% and 5.89%, respectively, compared to the control group. The AP and AK contents in the S-2 and M-2 groups showed some decrease compared to the control group. Research has shown that

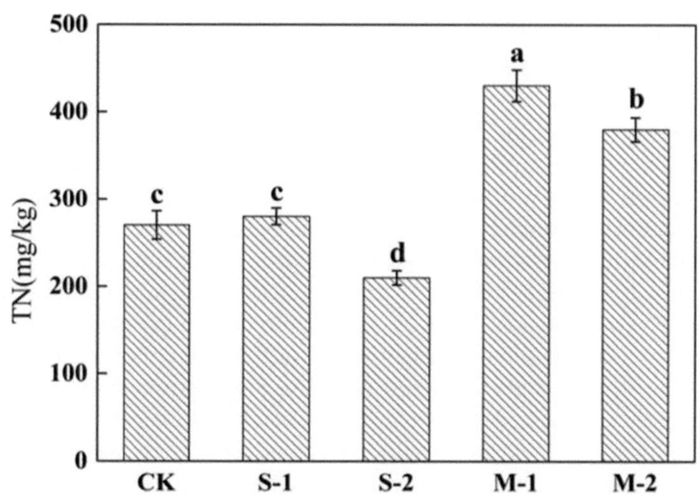

Fig. 5.1 Changes in soil total nitrogen under different fertilization methods. *Note* Different lowercase letters indicate significant differences between treatments at the 5% probability level

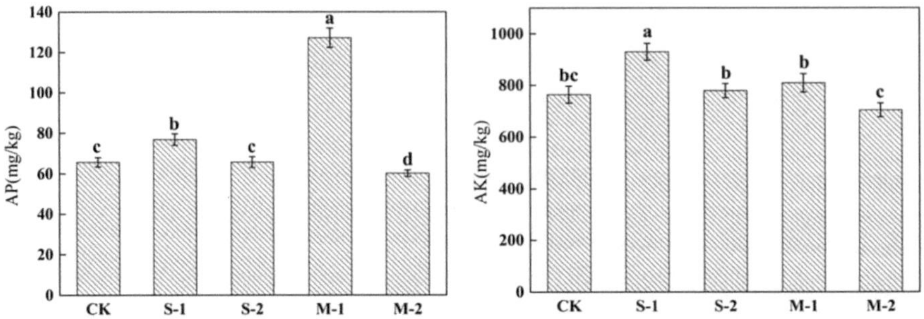

Fig. 5.2 Changes in soil available phosphorus and available potassium under different fertilization methods. *Note* Different lowercase letters indicate significant differences between treatments at the 5% probability level

applying large amounts of organic fertilizer may increase the downward migration of available phosphorus in the soil profile, thus posing a potential pollution risk to groundwater (Chardon et al. 1997). Therefore, in actual production, attention should be paid to the balanced supply of various nutrients to avoid the negative environmental effects caused by excessive and unbalanced fertilization.

5.1.3 Changes in Soil Organic Matter and Humic-Like Substances Under Different Fertilization Regimes

During the transformation and decomposition of organic matter (OM), energy and various nutrients that can be utilized by crops and microorganisms are released. As shown in Fig. 5.3, the OM content in the soil increased after fertilization. The most significant increase was in the M-2 group, which increased by 75.52% compared to the control group (34.93 g kg^{-1}). The S-1, S-2, and M-1 groups increased by 30.40%, 19.96%, and 40.60%, respectively, compared to the control group. The increases in S-1 and S-2 were not as pronounced as those in M-1 and M-2. The results indicate that fertilization is beneficial for increasing the organic matter content in the soil, and the combined application of organic fertilizer and a soil conditioner is more conducive to carbon sequestration in the soil. The findings of scholars who have studied the effects of organic fertilizer application on soil organic carbon and total carbon content are consistent with the results of this study. For example, Qi et al. found that the organic matter content increased most significantly in soils where organic fertilizer was applied (Qi 2016). The research by Liang et al. showed that the combined application of organic fertilizers can effectively increase the soil's total carbon and total organic carbon stocks, as well as the carbon sequestration rate (Liang et al. 2011). This is consistent with the results of the present study.

In this research, the soil organic matter content after fertilizer treatment was significantly higher than in the control group, and the effect of M-1 and M-2 on

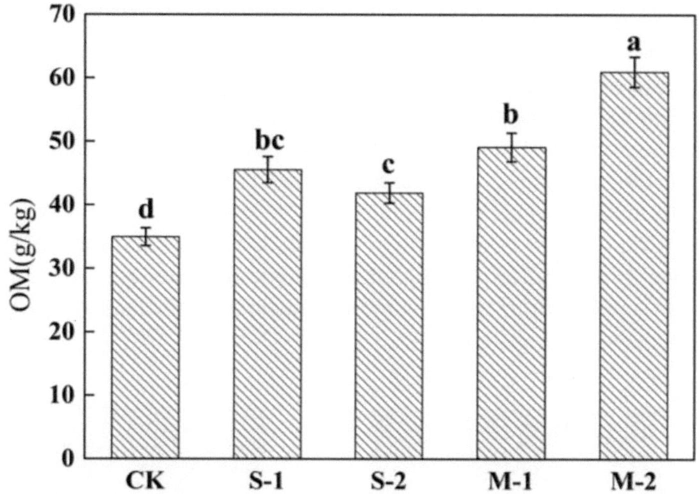

Fig. 5.3 Changes in soil organic matter under different fertilization methods. *Note* Different lowercase letters indicate significant differences between treatments at the 5% probability level

OM was more pronounced. The main reasons are: on one hand, BOF and Conditioner A contain a large amount of OM, which can directly increase the soil organic carbon content; on the other hand, the easily metabolizable small-molecule carbon contained in Conditioner A can provide effectively utilized nutrients for microorganisms, rapidly activating the microbes in the soil. The formation and transformation of soil organic matter are closely related to the abundance and diversity of the soil microbial community. An increase in microbial activity can promote the activities of microorganisms in transforming and decomposing OM. Additionally, soil microorganisms will synthesize and secrete certain decomposition-resistant organic substances, such as lipopolysaccharides and decane, in response to changes in the soil environment, which helps to promote the accumulation of soil organic matter. Finally, good soil aggregates have an encapsulating effect on organic matter, which can physically protect the internal organic matter from microbial decomposition, thereby increasing the stability of soil organic matter.

Humic substances are the main component of soil OM, possessing complex molecular structures, a high degree of molecular condensation and humification, and are not easily degraded, allowing for the sustained release of nutrients during long-term application. As shown in Fig. 5.4a, the three-dimensional fluorescence components of soil soluble organic matter are divided into five regions. Regions I, II, III, IV, and V represent tyrosine, tryptophan, fulvic acid, soluble microbial by-products, and humic acid, respectively. The corresponding Pi,n value for each region reflects the relative content of each substance. He et al. suggested that fluorescence peaks near the excitation/emission wavelength of 288/455 nm represent humic-like substances, and the change in peak intensity is directly proportional to the change in humus content (He et al. 2016). As shown in Fig. 5.4a, compared to CK, the fluorescence peak signal intensity in Region V was significantly enhanced in S-2, M-1, and M-2, while the signal intensity in Region II was enhanced in S-1. From the volume integration ratio of the three-dimensional fluorescence regions (Fig. 5.4b), the Pi,n of Region V in CK was 64.0%. The Pi,n of Region V in the M-1 and M-2 groups was significantly higher than in the other treatment groups, increasing to 72.1% and 74.8%, respectively, which is consistent with the changes in fluorescence intensity. This indicates that the degree of humification of organic matter in the soil increased after fertilization, and the application of organic fertilizer can promote the formation of humus in the soil, especially when organic fertilizer and soil conditioners are applied in combination, which is more beneficial for increasing the content of humic-like substances in the soil.

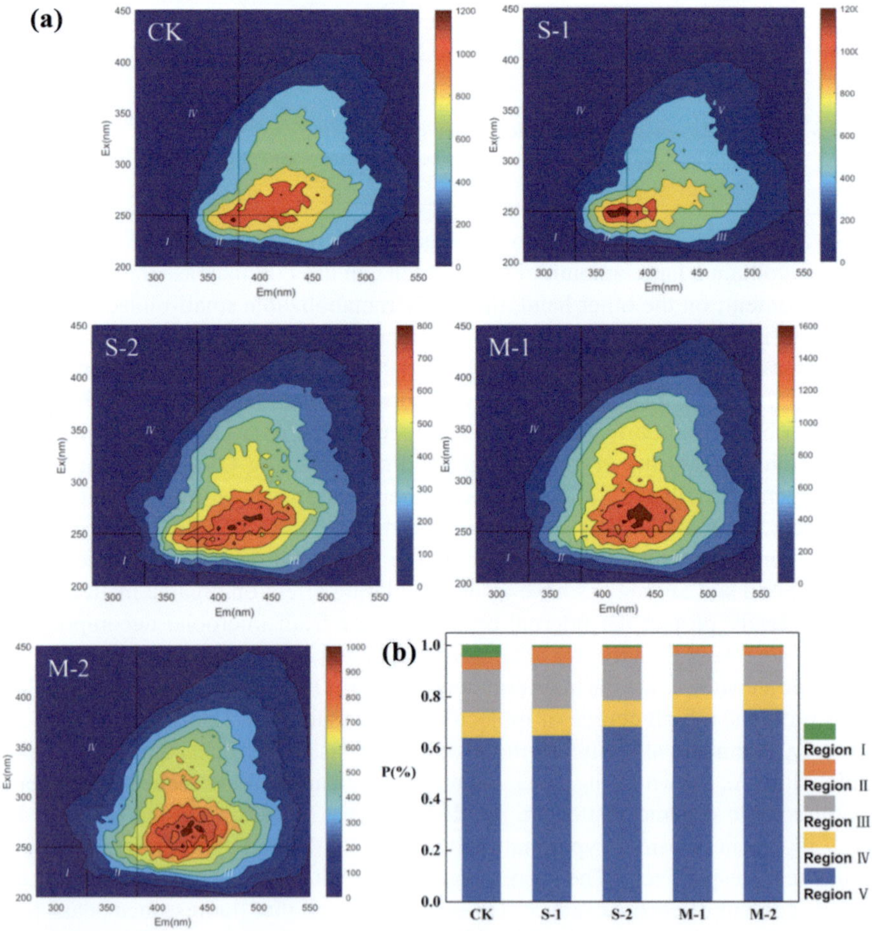

Fig. 5.4 **a** Changes in three-dimensional fluorescence spectra under different fertilization methods; **b** Volume integration ratio of three-dimensional fluorescence regions under different fertilization methods

5.1.4 Changes in Soil Aggregates Under Different Fertilization Regimes

Soil aggregates are the core part of soil structure, representing the basic structural units of soil formed through the action of biological and environmental factors. They play a crucial role in regulating soil properties and fertility, as well as improving ecological functions. Moreover, soil aggregates are important carriers of soil nutrients and are also key sites for the transformation and accumulation of organic matter. A good soil structure can provide sufficient oxygen, water, and nutrients for crop root growth, thereby ensuring crop quality and the sustainable development of the soil. Among these, aggregates with a particle size > 0.25 mm are generally referred to as soil macroaggregates, are positively correlated with soil fertility, and represent the best structure in terms of particle structure, stability, and fertility status in the soil. R0.25, Mean Weight Diameter (MWD), and Geometric Mean Diameter (GMD) can reflect the soil's fertility, the stability of its structure, and its resistance to erosion. The larger the R0.25, MWD, and GMD, the higher the degree of aggregation and the stronger the stability of the aggregates' average particle size. A stable soil aggregate structure, especially for macroaggregates, can prevent the leaching of colloidal components containing valuable nutrients and protect the organic matrix from microbial attack and erosion.

As shown in Fig. 5.5, in the control group, aggregates with a particle size of 0.25–2 mm accounted for the largest proportion. After fertilization, the proportion of aggregates with a particle size > 2 mm increased compared to the control group. Among these, the proportion of aggregates with a particle size > 5 mm increased significantly after fertilization compared to CK, while the proportion of aggregates with a particle size of 0.5–1 mm decreased significantly. Aggregates with a particle size of 2–5 mm have the strongest correlation with soil erosion. After fertilization, the proportion of large aggregates with a particle size of 2–5 mm increased, especially in M-1 and M-2, where the proportion of 2–5 mm aggregates was 2.24 and 2.41 times that of CK, respectively. The stability results for soil aggregates in the different treatment groups are shown in Table 5.5. The stability of soil aggregates improved in all fertilization treatment groups. The order of R0.25 for the different treatment groups was M-2 (69.78%) > M-1 (66.93%) > S-1 (60.34%) > S-2 (57.86%) > CK (57.39%). M-1 and M-2 significantly increased the MWD and GMD of the soil aggregates. The M-1 treatment group had the highest MWD (2.81 mm) and GMD (1.96 mm) values among all treatment groups, which were 1.46 and 1.58 times those of CK, respectively. The application of bio-organic fertilizer, especially in combination with a soil conditioner, can enhance soil aggregation, promote the cementation of small-sized aggregates into large-sized aggregates, thereby increasing the content of large aggregates and enhancing their formation and stability, which in turn promotes the optimization of soil structure.

The combined application of bio-organic fertilizer and the soil conditioner effectively increased the OM and humic-like substances in the soil. As important cementing agents, OM and humic-like substances can promote the aggregation of

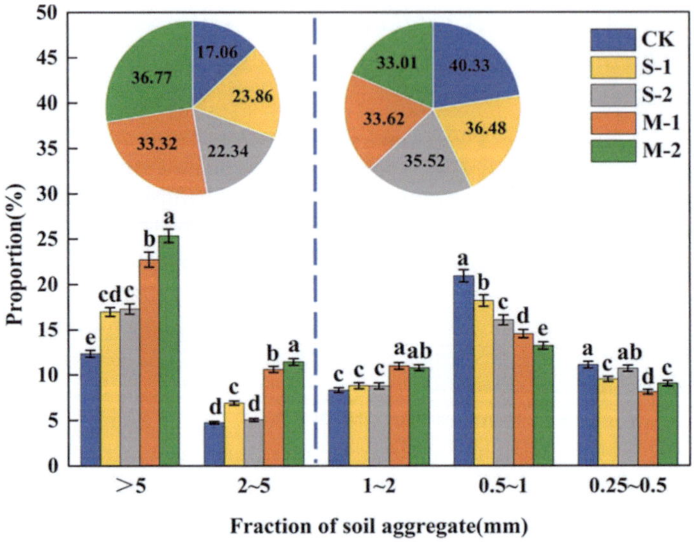

Fig. 5.5 Particle size distribution of water-stable aggregates in soil under different fertilization methods. *Note* Different lowercase letters indicate significant differences between treatments at the 5% probability level

Table 5.5 Stability of soil aggregates under different fertilization methods

	CK	S-1	S-2	M-1	M-2
$R_{0.25}$ (%)	57.39 ± 1.84b	60.34 ± 2.11b	57.86 ± 1.168b	66.93 ± 2.08a	69.78 ± 2.51a
MWD (mm)	1.93 ± 0.063d	2.31 ± 0.067b	2.30 ± 0.071bc	2.71 ± 0.079ab	2.81 ± 0.087a
GMD (mm)	1.24 ± 0.043d	1.51 ± 0.048b	1.48 ± 0.052bc	1.88 ± 0.058ab	1.96 ± 0.063a

Note Different lowercase letters indicate significant differences between treatments at the 5% probability level

soil from microaggregates to macroaggregates and are beneficial for promoting the stability of soil aggregates (Liu et al. 2019). Furthermore, the scientific combination of organic fertilizer and soil conditioner can improve the soil bacterial community and function, forming organo-mineral complexes by binding soil particles into soil aggregates through the release of exopolysaccharides, thereby improving soil structure. *Sphingomonadaceae* and *Xanthomonadaceae* are the main bacterial families whose abundance is positively correlated with macroaggregates. *Sphingomonadaceae* and *Xanthomonadaceae* can secrete extracellular polysaccharides, which are beneficial for promoting the formation and stability of aggregates. A large amount of Ca^{2+} can cause the soil to harden, and a high concentration of Na^{+} has a strong dispersive effect on soil particles, which will destroy the structure of aggregates and lead to hindered movement of gas and water. In the results, it was observed that the changes in the aggregates of the S-2 treatment group were caused by the increase in Ca^{2+} and Na^{+} content (Table 5.5).

5.1.5 Summary

(1) The combined application of bio-organic fertilizer and soil conditioner is beneficial for lowering soil pH, alleviating soil salinization, and reducing soil bulk density. However, the excessive application of organic fertilizer can raise soil pH. This indicates that a scientific approach to fertilizer application is more effective for adjusting soil acidity and alkalinity toward the optimal range for soil fertility utilization (pH 6.5–7.5), as well as for increasing soil porosity and permeability.
(2) Overall, the application of fertilizer did not significantly increase soil nitrogen, phosphorus, or potassium. For total nitrogen, the combined application of organic fertilizer and soil conditioner in groups M-1 and M-2 resulted in levels that were 1.59 and 1.42 times higher than the control group, respectively. For available phosphorus, only M-1 showed a significant increase, while S-1 was beneficial for increasing available potassium.
(3) Under the synergistic effect of bio-organic fertilizer and soil conditioner, the increase in organic matter was greater than with the application of organic fertilizer alone. The content of humic substances in the soil also increased significantly. This shows that the ability of organic fertilizer combined with a soil conditioner to enhance and stabilize soil organic matter is superior to that of applying organic fertilizer alone.
(4) Compared to the no-fertilizer control, the combined application of organic fertilizer and soil conditioner increased the proportion of soil aggregates > 2 mm in diameter and significantly improved the stability of these aggregates. This indicates that the combined application of bio-organic fertilizer and soil conditioner is advantageous for the retention of nutrients and the improvement of soil structure.

5.2 Effects of the Synergistic and High-Efficiency Soil Conditioner on the Migration and Transformation of Soil Salt

With an increasing number of planting years, excess fertilizer accumulates, and salts gradually build up in the cultivated layer. This leads to soil compaction and poor permeability, which affects the rhizosphere micro-ecological environment. Excessive salt accumulation in the soil can lower the osmotic potential of the soil solution in the crop rhizosphere, which in turn affects the ability of crop roots to absorb nutrients. This can cause physiological disorders in greenhouse crops and seriously impact the sustainable development of greenhouse vegetable production. The degree of salinization also affects the balanced supply of soil nutrients and their balanced absorption by crops. For example, excessive Na^+ can interfere with the absorption of ions such as K^+, Ca^{2+}, and Mg^{2+} by the crop. Ca^{2+} has a fixing effect on phosphorus, and an excess of Ca^{2+} can reduce the availability of phosphorus. This imbalanced absorption not only causes nutritional disorders in crops, which inhibits their growth, but it also has a single-salt toxic effect on the crops. Furthermore, secondary salinization in greenhouse soils can directly affect the activity of soil microorganisms and can also indirectly impact their living environment by altering some of the physical and chemical properties of the soil, leading to changes in the population, quantity, and activity of soil microorganisms in greenhouse soils.

5.2.1 Effects on Total Soluble Salts in Soil Under Different Fertilization Regimes

The total soluble salt content is a key physicochemical indicator of saline-alkali soil and is the most important factor affecting agricultural production. As shown in Fig. 5.6, S-1, M-1, and M-2 all reduced the salt content to some extent. Compared with the CK treatment (3.49 g kg^{-1}), the TDS content in the M-1 and M-2 groups decreased by 3.44% and 32.38%, respectively, with the reduction in M-2 being the most significant. However, the S-2 group (6.81 g kg^{-1}), which received a large amount of BOF alone, led to an increase in TDS content, which was 1.95 times that of the CK group. This is because the application of fertilizer in the S-2 treatment group increased the content of Ca^{2+} and Na^+ in the soil, which destroyed the soil aggregates and led to the accumulation of soil salts. In contrast, the M-1 and M-2 treatment groups increased the OM and humus in the soil, which improved the soil structure. OM can reduce capillary action, preventing water-soluble salts from evaporating to the surface. Moreover, a large amount of macromolecular humus plays a role in the fixation of ions. Humus is rich in oxygen-containing acidic functional groups and has strong ion exchange and complexation capabilities (Qi 2016), and it may complex with salt ions to reduce their effectiveness. Additionally, the formation and stability of soil aggregates can improve soil structure, increase soil permeability and porosity, and thereby inhibit the accumulation of salt in the surface soil. Furthermore, colloids in the aggregates can adsorb ions through van der Waals forces or electrostatic forces. Research has also shown that microorganisms play a key role in the improvement of saline-alkali soils (Schulten and Schnitzer1993). The growth and reproduction of microorganisms will absorb and fix inorganic nutrients in the soil and convert them into organic nutrients, thereby reducing the concentration and electrical conductivity of the soil solution.

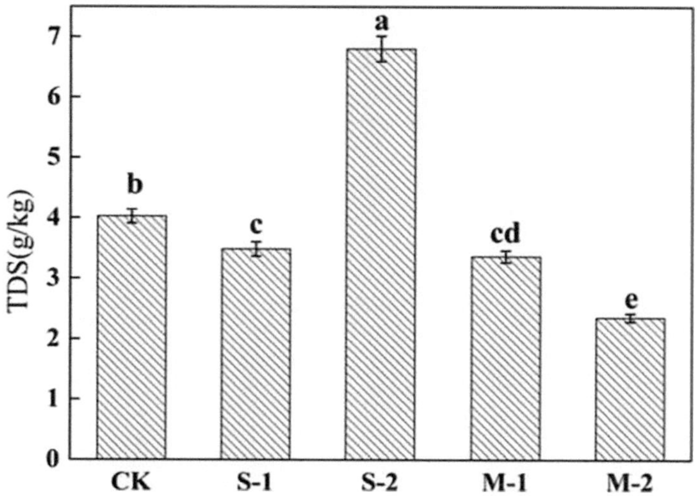

Fig. 5.6 The effect of different fertilization treatments on the total soluble salts in the soil. *Note* Different lowercase letters indicate significant differences between treatments at the 5% probability level

5.2.2 Effects on the Ionic Composition of Soil Salts Under Different Fertilization Regimes

Compared to coastal saline soils, the main cations in greenhouse soils are Ca^{2+}, K^+, and Na^+, while the main anions are NO_3^- and Cl^-. This is a primary characteristic of secondary salinization in greenhouse soils. A large amount of Ca^{2+} can make the soil alkaline and lead to soil compaction. High concentrations of Na^+ have a strong dispersive effect on soil particles. Both of these can destroy the aggregate structure, and once the aggregates are damaged, the flow of gas and water is hindered.

In the S-1, M-1, and M-2 soils, the ratio of anions to cations was approximately 38:61, which is balanced compared to the 6:94 ratio in the CK group (Fig. 5.7a). With the exception of S-2, the application of fertilizer, especially the combined application of organic fertilizer and conditioner, can reduce the percentage of cations and increase the proportion of anions in the soil. This helps to regulate the ionic balance in the soil, creating a healthy soil environment and avoiding single-salt toxicity to crops and microorganisms.

As can be seen from Fig. 5.7b, with the exception of S-2, the other three fertilization treatment groups significantly reduced the concentrations of Ca^{2+}, Na^+, and K^+ in the soil. Among these, M-2 had the most significant effect on reducing cation content. Compared to the control group, K^+ (225.41 mg kg^{-1}), Na^+ (482.72 mg kg^{-1}), and Ca^{2+} (590.40 mg kg^{-1}) decreased by 66.1%, 53.9%, and 51.8%, respectively. The proportions of K^+, Na^+, and Ca^{2+} in the total ions also decreased from 21.59%, 30.07%, and 35.13% to 10.82%, 20.45%, and 25.02%, respectively. Fertilizer application is beneficial to crop growth, which in turn promotes the absorption of soil salts

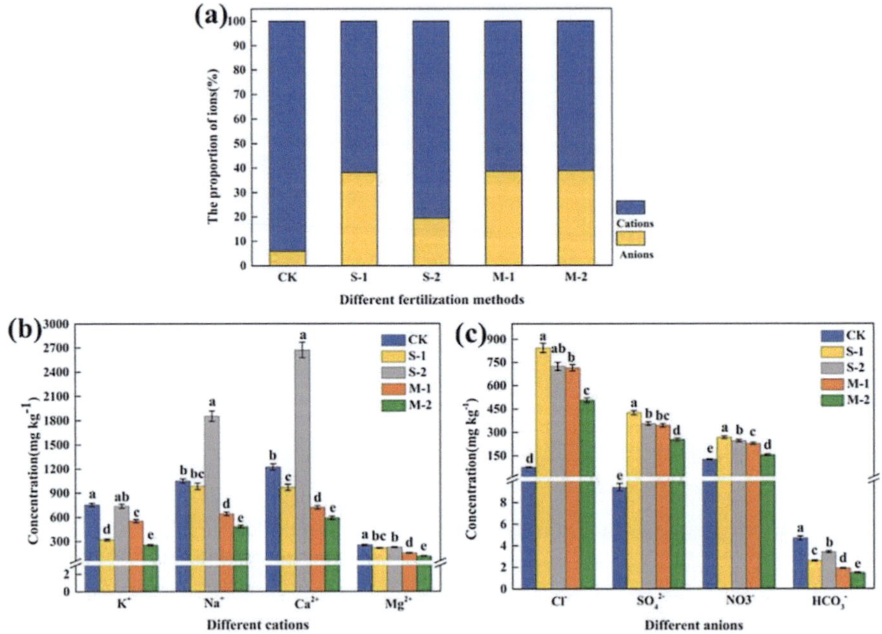

Fig. 5.7 **a** The ratio of anions to cations in the soil under different fertilization treatments; **b** The effect of different fertilization regimes on salt cations in the soil; **c** The effect of different fertilization regimes on salt anions in the soil. *Note* Different lowercase letters indicate significant differences between treatments at the 5% probability level

by the crops. According to Tattini et al., some non-halophytes, such as radishes, are salt-tolerant and can accumulate large amounts of K^+ in their tissues (Tattini et al. 2010). Compared to the control group, the contents of Cl^-, SO_4^{2-}, and NO_3^- in the soil increased after fertilization (Fig. 5.7c). The fertilizers all contained Cl^- and SO_4^{2-}, which directly increased the content of these ions in the soil. The increase in NO_3^- content may be due to the stimulation of microbial activity by the fertilizer application, which enhanced the activity of nitrifying bacteria and promoted ammonification and nitrification. The fertilizer in M-2 contained a large amount of organic matter, which can reduce the activity of polyvalent cations such as Ca^{2+} through complexation. In addition, the fertilizer contained a large amount of macromolecular humus, which plays an important role in the fixation of ions. On the one hand, humic acid is rich in oxygen-containing acidic functional groups and has strong ion exchange and complexation capabilities, which reduces the amount of cations available for exchange. On the other hand, humic acid is an organic colloid that adsorbs cations through Van der Waals forces and electrostatic forces.

5.2.3 Effects on the Migration and Transformation of Soil Salt Ions Under Different Fertilization Regimes

To clarify the effect of different fertilization regimes on the migration and transformation of ions in the soil, five sets of simulation experiments were designed to study the changes in the concentration of major salt ions in the soil over time. The five experimental groups were: no fertilizer (CK), single application of organic fertilizer (T1), single application of soil conditioner A (T2), combined application of organic fertilizer and conditioner A (T3), and combined application of organic fertilizer and conditioners A and B (T4). The fertilization regimes and application rates for the simulation experiments are described in the experimental design for the simulation experiments.

As shown in Fig. 5.8, excessive Na^+ remaining in the soil can disperse the soil structure. After 1 day of fertilization, the concentration of Na^+ in the soil increased in all groups, with the largest increase in the T2 treatment group, followed by T3 and T4. After 28 days of fertilization, the Na^+ concentration in the T2 treatment group gradually decreased. The Na^+ concentration fluctuated little during the experiment in the CK, T1, and T3 groups. In CK and T3, the Na^+ concentration slowly decreased after 1 day, while in T1, the Na^+ concentration slowly increased after 1 day of fertilization. At the end of the experiment, the Na^+ concentration after fertilization was higher than that in the soil before fertilization in all treatment groups, especially in the group where soil conditioner A was applied. This is because soil conditioner A was prepared from food waste, which has a high salt content. The resulting soil conditioner also contained a relatively high amount of salt, which directly added Na^+ to the soil.

K^+ is a nutrient required by crops and is also a base cation in the soil. The trends of K^+ concentration changes in the CK and T1 treatment groups were similar, reaching a maximum value at 14 days. After 14 days, the K^+ concentration gradually decreased. In the CK treatment group, a large amount of K^+ was lost with the water, and the K^+ concentration was lower than the concentration before treatment. The trends of K^+ concentration changes in the T2, T3, and T4 treatment groups were similar. The K^+ concentration increased 1 day after fertilization and then slowly decreased.

Mg^{2+} is also a major base cation in the soil. After fertilization, the Mg^{2+} concentration in the soil gradually increased and reached its maximum value on the 7th day. After 7 days, the Mg^{2+} concentration in T1 gradually decreased, while the Mg^{2+} concentration in T2, T3, and T4 fluctuated, with an overall decreasing trend. At the end of the experiment, the Mg^{2+} concentration in T1 and T2 tended toward the Mg^{2+} concentration in the soil before fertilization. In the unfertilized CK treatment group, the Mg^{2+} concentration gradually decreased, and at the end of the experiment, it was much lower than the Mg^{2+} concentration in the fertilized treatment groups.

Some studies have shown that Ca^{2+} is the main base cation in greenhouse soils, and its concentration is very different from that in open-field soils, being about 2.7 times higher (Yu 2006). In this study, Ca^{2+} was also the most abundant cation in the soil. The Ca^{2+} concentration gradually increased within 1 to 7 days after fertilization.

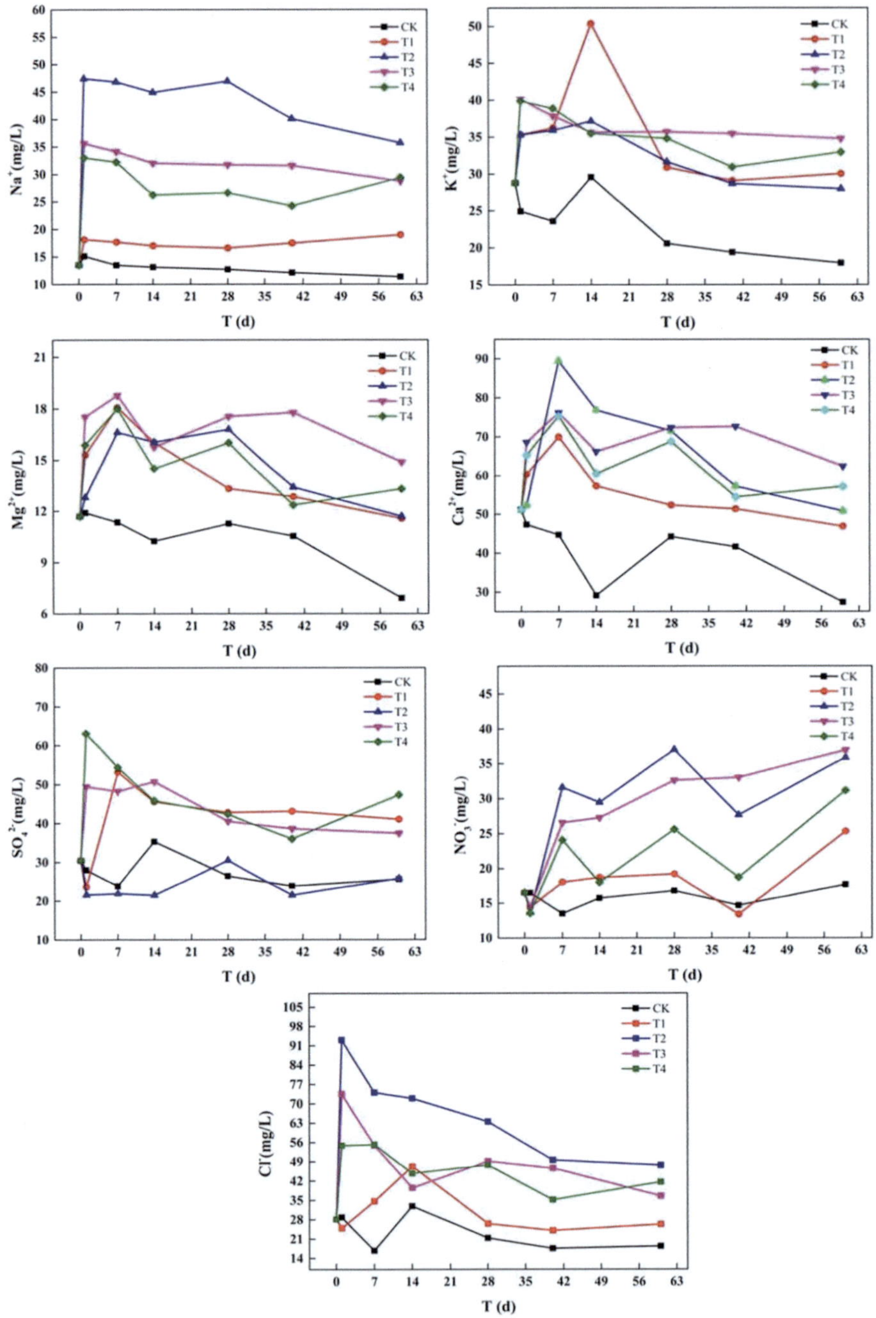

Fig. 5.8 Effects of different fertilization treatments on the changes of salt ions in the soil

After 7 days, the Ca^{2+} concentration in T1 and T2 gradually decreased, while the Ca^{2+} concentration in T3 and T4 fluctuated downwards. In the unfertilized CK treatment group, the Ca^{2+} concentration decreased from day 1 to day 14, increased from day 14 to day 28, and then gradually decreased after 28 days. At the end of the experiment, a large amount of Ca^{2+} was lost in the CK group, and its concentration was much lower than that in the soil before treatment.

For SO_4^{2-}, the trends in T1, T3, and T4 were similar. The SO_4^{2-} concentration in T4 reached its highest value 1 day after fertilization. From day 1 to day 40, the SO_4^{2-} concentration gradually decreased. From day 42 to day 60, the SO_4^{2-} concentration gradually increased. The SO_4^{2-} concentration in T1 and T3 gradually decreased after 7 days. In the later stage of the experiment, the concentration changes in these three treatment groups tended to be stable, and the range of change decreased. In CK and T2, the SO_4^{2-} concentration decreased in the early stage of the experiment and then slowly fluctuated toward the initial value in the later stage.

For NO_3^-, the concentration in all treatment groups showed an upward trend. The NO_3^- concentration in the soil decreased 1 day after fertilization. Upon testing, the nitrite ion content in the soil was found to have increased. After 1 day, the NO_3^- concentration in T3 gradually increased. The trends of NO_3^- concentration changes in T2 and T4 were similar, with the concentration increasing from day 1 to day 7, decreasing from day 7 to day 14, increasing from day 14 to day 28, decreasing from day 28 to day 40, and then increasing again from day 40 to day 60. The trend of change in T1 after 28 days was similar to that in T2 and T4. Throughout the experiment, the NO_3^- concentration in T1, T2, and T4 fluctuated greatly.

Cl^- is chemically stable in the soil and is not easily adsorbed by soil colloids, but it moves with the flow of water. One day after fertilization, the Cl^- concentration in T2, T3, and T4 reached its highest value and then gradually decreased. The trends of Cl^- concentration changes in CK and T1 were similar. The Cl^- concentration increased from day 7 to day 14, reaching its maximum value at 14 days, and then gradually decreased after 14 days.

The magnitude of change in salt ions in the simulation experiment was different from that in the field experiment. The reduction in salt ions in the simulation experiment was not large, possibly because no crops were planted in the simulation. Crops have a certain absorption and transformation effect on soil salts. Through the addition of fertilizer to the soil, some of the salt ions are leached out by irrigation water under a good soil structure, some are metabolized and transformed by microorganisms, and some are adsorbed and fixed by organic matter and humus in the soil. After fertilization, the utilization rate of ions in the soil increases, and the excess salt ions are leached out with the irrigation water. This indicates that fertilization is beneficial to the transformation and utilization of salt ions in the soil, and it also promotes the migration of ions in the soil, ultimately reducing the accumulation of salt ions in the surface layer of the soil.

5.2.4 Summary

(1) A moderate application of organic fertilizer (S-1) and the combined application of organic fertilizer with soil conditioners (M-1, M-2) were all beneficial for reducing soil salinity. The salt reduction effect of M-2 was the most significant, with a 32.38% decrease compared to CK. Conversely, the application of a large amount of organic fertilizer (S-2) increased the soil salt concentration. This indicates that only a scientific approach to fertilization is conducive to lowering soil salinity and promoting healthy soil development.
(2) The combined application of organic fertilizer and soil conditioner is beneficial for reducing the concentrations of K^+, Na^+, and Ca^{2+} ions in the soil. Compared to the 6:94 ratio in the CK group, the anion-to-cation ratios in M-1 and M-2 were approximately 38:51. This is advantageous for regulating the balance of anions and cations in the soil and reducing the toxic effects of single salts.
(3) The organic fertilizer and soil conditioner had a significant impact on Na^+, K^+, Mg^{2+}, Ca^{2+}, SO_4^{2-}, and Cl^- in the soil. The ions introduced with the fertilizers are absorbed, adsorbed, transformed, and leached over a period of time and will not cause an accumulation or a large loss of soil salt ions. In contrast, not applying fertilizer leads to a significant loss of ions.

5.3 Relationship Between the Synergistic and High-Efficiency Soil Conditioner and the Soil Microbial Community

Soil contains an abundance of microorganisms that form soil microbial communities and respond to micro-environmental changes within the soil, playing a vital role in ecosystem productivity and biogeochemistry. Soil microorganisms interact with soil properties (such as porosity and soil aggregates) to transmit relevant information, and they adapt to changes in the soil environment by altering a series of biochemical processes. These organisms, often subjected to soil salinization, can tolerate or adapt to salt stress. They absorb and fix inorganic nutrients in the soil and convert them into organic nutrients, thereby reducing the concentration and electrical conductivity of the soil solution. The formation and transformation of soil organic matter are closely related to the abundance and diversity of the soil microbial community. In addition to cycling nutrients and immobilizing heavy metals, microorganisms can also promote plant growth. Of course, microbial byproducts are also essential binding agents that stabilize soil aggregates and increase the content of soil organic matter.

5.3.1 Microbial Community Composition of Different Fertilizers

As shown in Fig. 5.9a, the main microbial phyla in the bio-organic fertilizer were *Chloroflexi*, *Actinobacteria*, *Bacteroidetes*, and *Proteobacteria*, with relative abundances of 58.16%, 15.10%, 13.28%, and 7.12%, respectively. In conditioner A, the dominant phyla were *Firmicutes* and *Actinobacteria*, with relative abundances of 65.16% and 33.26%, respectively. In conditioner B, the main microbial phyla were *Actinobacteria*, *Firmicutes*, *Chloroflexi*, and *Proteobacteria*, with relative abundances of 62.93%, 19.57%, 5.37%, and 4.32%, respectively. In the bio-organic fertilizer, the phylum *Chloroflexi* was mainly composed of the classes *Anaerolineae* and *Thermomicrobia*, with relative abundances of 42.48% and 14.65%, respectively (Fig. 5.9b). The class *Flavobacteriia*, belonging to the phylum *Bacteroidetes*, had a relative abundance of 9.54%. In soil conditioner A, the most abundant class was *Bacilli*, with a relative abundance of 65.07%. In soil conditioner B, the most abundant class was *Actinobacteria*, followed by *Bacilli*.

The primary bacteria in the bio-organic fertilizer belong to the phylum *Chloroflexi*, which can participate in the degradation of plant-derived compounds such as cellulose (Wang et al. 2018). The main microorganisms in soil conditioner A are from the classes *Bacilli* and *Actinobacteria*, while in soil conditioner B, the main microorganisms are from the classes *Actinobacteria* and *Bacilli*. The class *Bacilli* has good adaptability and can degrade large molecular substances in the soil, such as cellulose, starch, and protein, into smaller molecules. It also plays a major role in nitrogen fixation. The class *Actinobacteria* is widely present in soil environments, and its relative abundance shows a significant positive or negative correlation with pH. It is often involved in the transformation of OM, can decompose recalcitrant carbon-based compounds, and can participate in nitrogen fixation. There were no significant differences in the relative abundance of fungi at the phylum level among the different fertilizers. The relative abundance at the genus level is shown in Fig. 5.9c. The main fungi in the bio-organic fertilizer were *unclassified_p_Ascomycota*, *Acremonium*, and *unclassified_k_Fungi*. The main fungi in conditioner A were *unclassified_c_Eurotiomycetes*, *Xerochrysium*, and *Aspergillus*. The main fungi in the conditioner B were *Microascus* and *unclassified_p_Ascomycota*.

5.3.2 Analysis of Soil Microbial Diversity Under Different Fertilization Regimes

After sequence filtering, 106,038 valid sequences were obtained from the 5 soil samples. The coverage for all samples was higher than 98%, indicating that the sequencing depth was sufficient to truly reflect the microbial diversity in the samples. Based on the Shannon, Ace, Chao1, and Coverage indices, the effect of fertilization on bacterial abundance and diversity was not significant (Table 5.6).

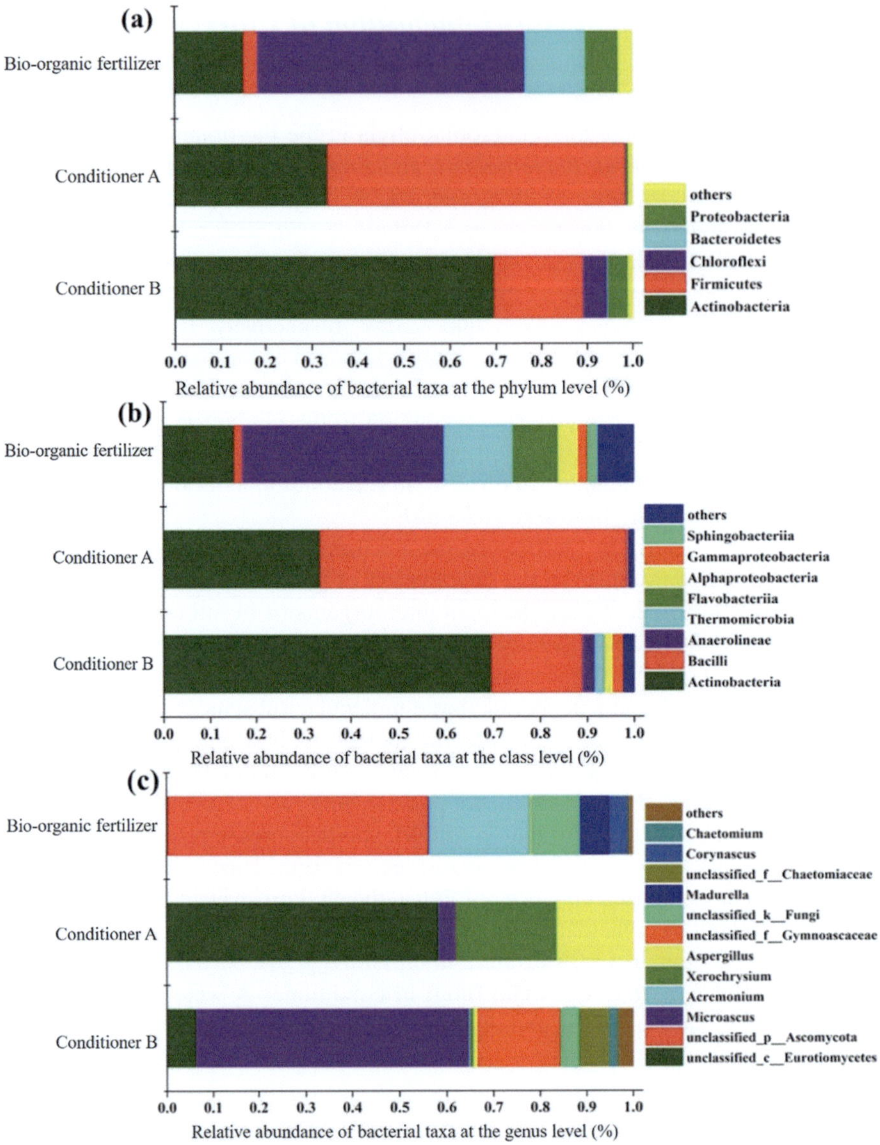

Fig. 5.9 **a** Bacterial communities of different fertilizers at the phylum level; **b** Bacterial communities of different fertilizers at the class level; **c** Fungal communities of different fertilizers at the genus level

Table 5.6 Effects of different fertilization treatments on bacterial richness and diversity

	Valid sequences	Shannon	Ace	Chao1	Coverage
CK	54,416 ± 1197.15	6.317 ± 0.114	2278.92 ± 38.74	2329.46 ± 51.25	0.989 ± 0.016
S-1	59,234 ± 1184.68	6.488 ± 0.097	2293.63 ± 34.40	2346.89 ± 56.33	0.991 ± 0.0139
S-2	53,681 ± 1019.94	6.432 ± 0.116	2187.68 ± 37.19	2266.73 ± 43.07	0.990 ± 0.0158
M-1	71,355 ± 1284.39	6.293 ± 0.132	2316.25 ± 41.69	2347.38 ± 46.95	0.991 ± 0.0129
M-2	50,578 ± 859.83	6.146 ± 0.135	2294.16 ± 36.71	2322.96 ± 34.84	0.986 ± 0.0158

5.3.3 Effects of Different Fertilization Regimes on Soil Community Composition

Figure 5.10 shows the cluster analysis of microorganisms and the relative abundance of dominant bacterial phyla in the five different treatment groups. Hierarchical cluster analysis of bacterial community similarity indicates that the bacterial communities of the S-1 and S-2 soils have high similarity to the CK bacterial community, but not to the bacterial communities of M-1 and M-2 (Fig. 5.10a). This suggests that the soil conditioner has a significant impact on the bacterial community structure.

The dominant phyla observed across all treatments included *Proteobacteria, Acidobacteria, Bacteroidetes, Chloroflexi, Gemmatimonadetes, Actinobacteria*, and *Firmicutes.* In all five treatment groups, these dominant phyla accounted for over 92% of the bacterial sequences (Fig. 5.10b). Compared to the control soil, the fertilization treatments increased the relative abundance of *Proteobacteria, Bacteroidetes, Gemmatimonadetes*, and *Firmicutes*, and decreased the relative abundance of *Acidobacteria.* Among these, *Proteobacteria* and *Acidobacteria* were the two most sensitive phyla under the different treatments.

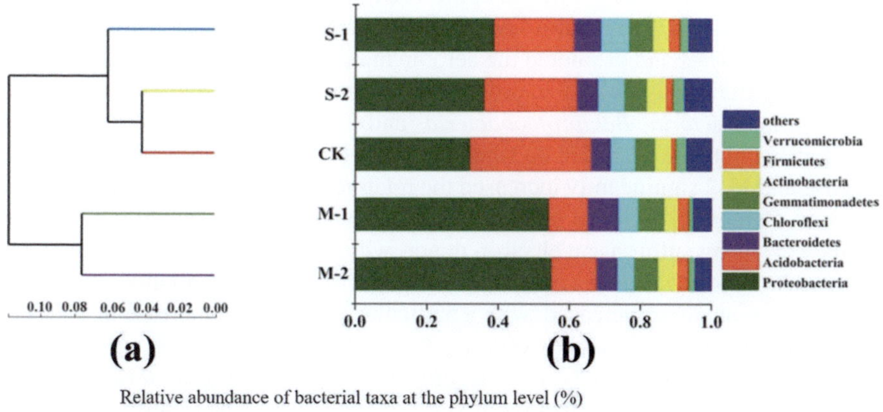

Fig. 5.10 **a** Hierarchical cluster analysis diagram of microorganisms; **b** Changes in the microbial community at the phylum level under different fertilization treatments

After fertilization, the relative abundance of *Proteobacteria* in the soil increased. M-1 and M-2, in particular, significantly increased the relative abundance of *Proteobacteria* from 31.97% in the control group to 54.20% and 55.00%, respectively. *Proteobacteria* are copiotrophic, colonizing nutrient-rich environments, and their increased abundance indicates an improvement in soil quality. After fertilization, the relative abundance of *Acidobacteria* in the soil decreased. The relative abundance in the M-1 and M-2 treatment groups decreased from 33.99% in the control group to 10.93% and 12.67%, respectively. *Acidobacteria* are adapted to survive in acidic environments and low-nutrient soil conditions and can be used as an indicator of soil infertility. In the M-1 and M-2 soils, the copiotrophic bacteria *Proteobacteria* gradually replaced the oligotrophic bacteria *Acidobacteria* as the dominant phylum, indicating that the soil quality improved after the M-1 and M-2 treatments. Compared to CK, S-1, and S-2, the relative abundances of *Chloroflexi* and *Bacteroidetes* were lower in the M-1 and M-2 soils. This reduction may be due to the relatively weaker competitiveness of these bacteria against other bacteria under high-nutrient conditions. Furthermore, the relative abundance of *Gemmatimonadetes* in the M-1 and M-2 soils was significantly higher than in the control soil.

Soil conditioner A contains small molecular available carbon that can provide directly absorbable and utilizable nutrients for microorganisms, increasing microbial abundance and activating dominant phyla. The stable aggregate structure and abundant nutrients in M-1 and M-2 provided a healthy and favorable rhizosphere micro-ecological environment for microorganisms, promoting the growth and reproduction of dominant phyla. The gradual replacement of oligotrophic bacteria (*Acidobacteria*) by copiotrophic bacteria (*Proteobacteria*) as the dominant phylum in M-1 and M-2 soils indicates that these treatments improved soil fertility, providing microorganisms with a large amount of directly absorbable and utilizable nutrients. Furthermore, the increased relative abundance of *Gemmatimonadetes* in M-1 and M-2 soils suggests that the combined application of bio-organic fertilizer and soil conditioner improved the physical properties of the soil and optimized its structure, leading to good permeability. This, in turn, provided a favorable soil micro-ecological environment for *Gemmatimonadetes* and promoted the aerobic metabolism of microorganisms.

The four main classes of *Proteobacteria* in the soil are *Alphaproteobacteria*, *Gammaproteobacteria*, *Betaproteobacteria*, and *Deltaproteobacteria* (Fig. 5.11). After fertilization, the relative abundance of all four *Proteobacteria* classes increased, and M-1 and M-2 significantly increased the relative abundance of *Alphaproteobacteria* and *Gammaproteobacteria*. Some bacteria within the *Gammaproteobacteria* can fix CO_2 using sulfur as an electron donor, which may be one of the reasons for the increase in soil OM after fertilization. After fertilization, the relative abundance of *Gemmatimonadetes* increased, with a particularly significant increase in M-1 and M-2, rising from 5.55% in the control group to 7.33% and 6.67%, respectively. However, fertilization did not have a significant effect on *Actinobacteria*, which is often involved in the transformation of soil organic matter. The relative abundance of the class *Bacilli* in the soil increased after fertilization, especially in M-1 and M-2, rising from 1.41% in the control group to 3.18% and 3.27%, respectively. It is well known that the function of *Bacilli* is to break down macromolecular substances such

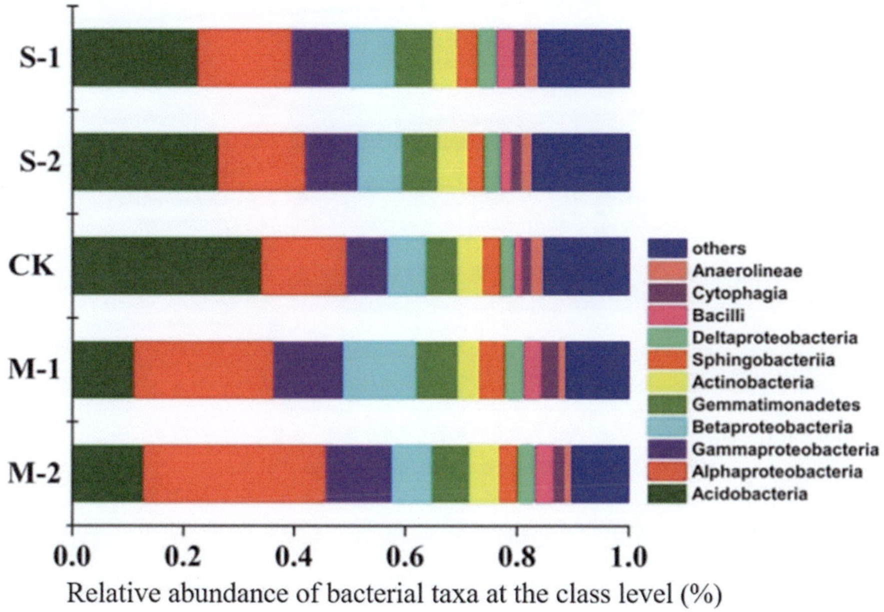

Fig. 5.11 Changes in the microbial community at the class level under different fertilization treatments

as cellulose, starch, and protein into small molecular substances (Hou et al. 2017). The increase in the relative abundance of *Bacilli* also indicates an increased content of macromolecular substances in the soils treated with M-1 and M-2. Furthermore, *Bacilli* in the soil play a dominant role in the nitrogen fixation process, which helps to reduce nitrogen loss from the soil.

As shown in Fig. 5.12, the major families changed significantly under different fertilization conditions. After fertilization, the relative abundance of *norank_c_Acidobacteria* in the soil decreased significantly, especially in the M-1 and M-2 treatment groups, where it fell to 4.52% and 6.06%, respectively, compared to the control group (12.18%). The relative abundance of *Blastocatellaceae_Subgroup_4* also decreased after fertilization, with the relative abundances in M-1 and M-2 decreasing from 10.83% in the control group to 2.75% and 3.78%, respectively. Compared to the control group, the relative abundances of the families *Sphingomonadaceae*, *Gemmatimonadaceae*, *Xanthomonadaceae*, and *Bacillaceae* increased in the soils of the M-1 and M-2 treatment groups. Among these, the relative abundance of *Sphingomonadaceae* increased from 7.23% in the control group to 11.00% and 15.95%, respectively; the relative abundance of *Gemmatimonadaceae* increased from 3.10% in the control group to 5.15% and 4.27%, respectively; and *Xanthomonadaceae* increased from 2.85% in the control group to 5.02% and 4.59%. *Sphingomonadaceae* and *Xanthomonadaceae* can secrete extracellular polysaccharides, which are beneficial for promoting the formation and stability of aggregates. *Xanthomonadaceae* is a

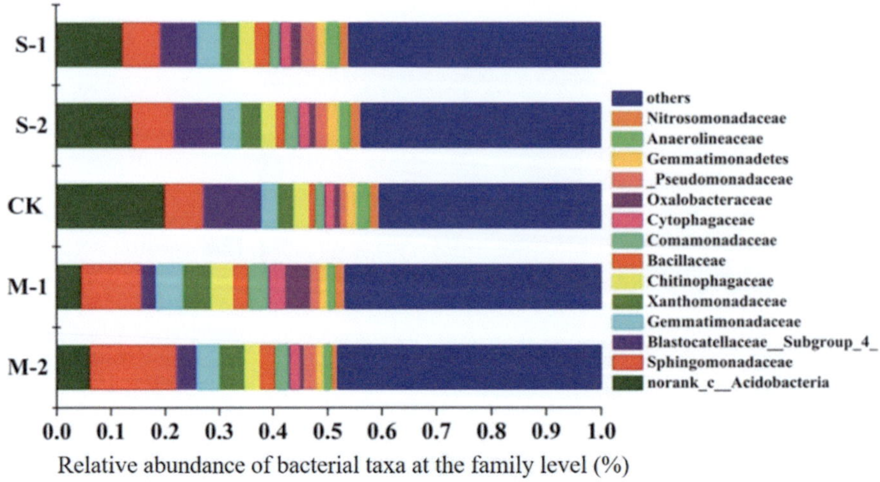

Fig. 5.12 Changes in the microbial community at the family level under different fertilization treatments

strictly aerobic bacterium that plays an important role in the decomposition of cellulose. The increase in the relative abundance of strictly aerobic bacteria also indicates an improvement in soil structure and an increase in soil permeability. It has been reported that *Sphingomonadaceae* is suitable for growth and reproduction in neutral pH soils and can participate in the decomposition and mineralization of recalcitrant substances (Zhang et al. 2017).

5.3.4 Correlation Analysis Between Soil Properties and Soil Microbes

The scientific combined application of organic fertilizer and soil conditioner can improve the soil bacterial community and its functions. It can also improve soil structure by causing soil particles to bind into aggregates through the release of extracellular polysaccharides, forming organic-mineral complexes. Organic matter and humus can combine with soil microaggregates to form large soil aggregates and improve the structure of greenhouse soil. The formation and stability of soil aggregates can increase soil porosity, thereby reducing the accumulation of salt in the surface soil (Liu et al. 2019). Research has also shown that microorganisms play a key role in the amelioration of saline-alkali soils, and that the organic matter and aggregates of all soils affect the soil microbial community (Mbarki et al. 2017). Therefore, we conducted a correlation analysis of soil organic matter and humus content, the proportion of aggregates > 2 mm in diameter, and total soluble salts to study the response relationship between soil properties and microorganisms. For the microorganisms, the top 13 most abundant families with a relative abundance greater than 0.1% were selected.

Figure 5.13 shows the relationship between soil properties and the bacterial community structure. As can be seen, SOM is positively correlated with humus and large aggregates. Humus is a part of soil organic matter, so an increase in organic matter also leads to an increase in humus. A higher content of organic matter and humus in the soil can promote the formation and stability of aggregates, while large aggregates provide better physical protection for organic matter and humus. Therefore, these three are closely linked and mutually reinforcing. OM, humus, and aggregates are all negatively correlated with TDS. The molecular structure of macromolecular humus is rich in oxygen-containing acidic functional groups, giving it strong ion exchange and complexation capabilities, which allow it to complex with salt ions and reduce their effectiveness. Stable large aggregates can improve soil structure and increase soil permeability. Organic matter in the soil can reduce capillary action, making it difficult for salts dissolved in water to be brought to the surface through evaporation. Additionally, colloids within the aggregates can adsorb electrons through Van der Waals forces or electrostatic forces.

The families *Sphingomonadaceae* and *Xanthomonadaceae* showed a significant negative correlation with TDS and a positive correlation with OM, humus, and large aggregates. Conversely, *Blastocatellaceae_Subgroup_4_* and *norank_c_Acidobacteria* showed a significant negative correlation with SOM, humus, and large aggregates, and a positive correlation with TDS. It has been reported that the family *Sphingomonadaceae* can adapt to a neutral pH and successfully compete with native organisms in various environments (Zhang et al. 2017). The families *Sphingomonadaceae* and *Xanthomonadaceae* can secrete extracellular polysaccharides, which are beneficial for promoting the formation and stability of aggregates. In contrast, *norank_c_Acidobacteria* is adapted to survive in acidic environments and low-nutrient soil conditions and can be used as an indicator of soil infertility. The scientific combined application of organic fertilizer and soil conditioner weakened the toxic effects of salinity, increased nutrient-rich organic matter, formed a good soil aggregate structure, and consequently increased soil permeability, which ultimately favored the growth and reproduction of dominant microbial groups.

5.3.5 Prediction of Soil Microbial Functions

As shown in Fig. 5.14, there were no significant differences in the abundance of different functional microorganisms among the different treatment groups. The main functions of microorganisms in the soil were protein turnover; amino acid transport and metabolism; lipid transport and metabolism; coenzyme transport and metabolism; carbohydrate transport and metabolism; synthesis and catabolism of secondary metabolites; intracellular transport and secretion; transport and metabolism of inorganic ions; and energy production and conversion.

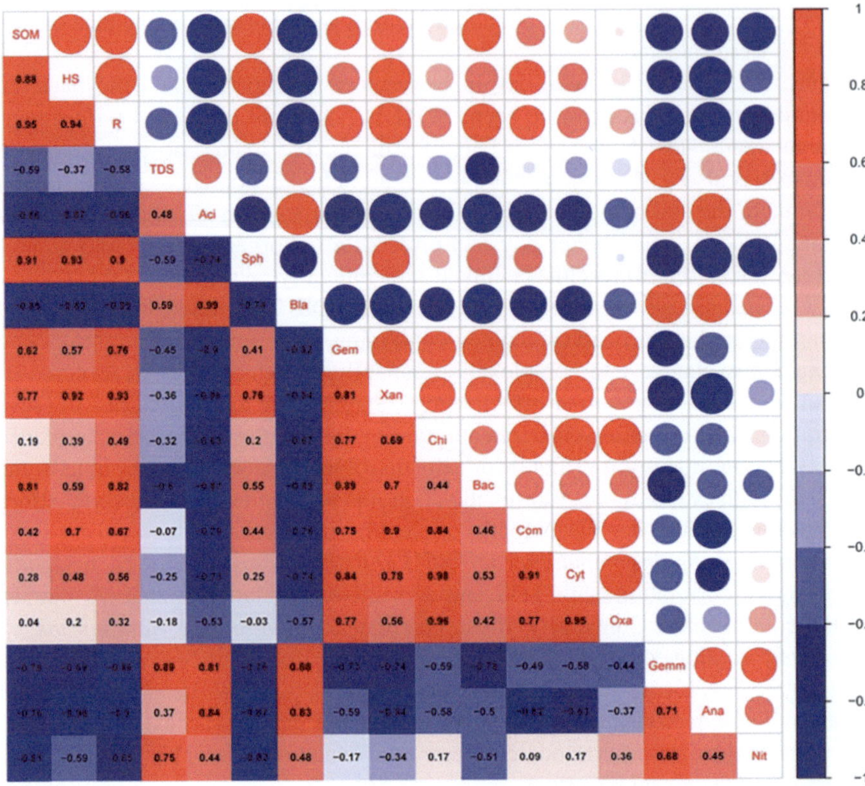

Fig. 5.13 Response between soil properties and the microbial community. SOM: soil organic matter; HS: humic substances; R: content of aggregates > 2 mm in diameter; TDS: total soluble salts; Aci, Sph, Bla, Gem, Xan, Chi, Bac, Com, Cyt, Oxa, Gemm, Ana, Nit are *respectively norank_c__Acidobacteria, Sphingomonadaceae, Blastocatellaceae__Subgroup_4_, Gemmatimonadaceae, Xanthomonadaceae, Chitinophagaceae, Bacillaceae, Comamonadaceae, Cytophagaceae, Oxalobacteraceae, Pseudomonadaceae, Gemmatimonadetes, Anaerolineaceae, Nitrosomonadaceae*

5.3.6 Summary

(1) The fertilization treatments did not have a significant effect on the diversity, richness, or functional structure of soil microorganisms.
(2) The organic fertilizer and soil conditioner had a relatively large impact on the soil microbial community. The microorganisms in M-1 and M-2 showed high similarity, while there was similarity among CK, S-1, and S-2. The synergistic application of organic fertilizer and soil conditioner can increase the abundance of dominant copiotrophic microbial groups while reducing the oligotrophic microbial communities in the soil. It can also increase the abundance of microorganisms that decompose macromolecules like cellulose, starch, and

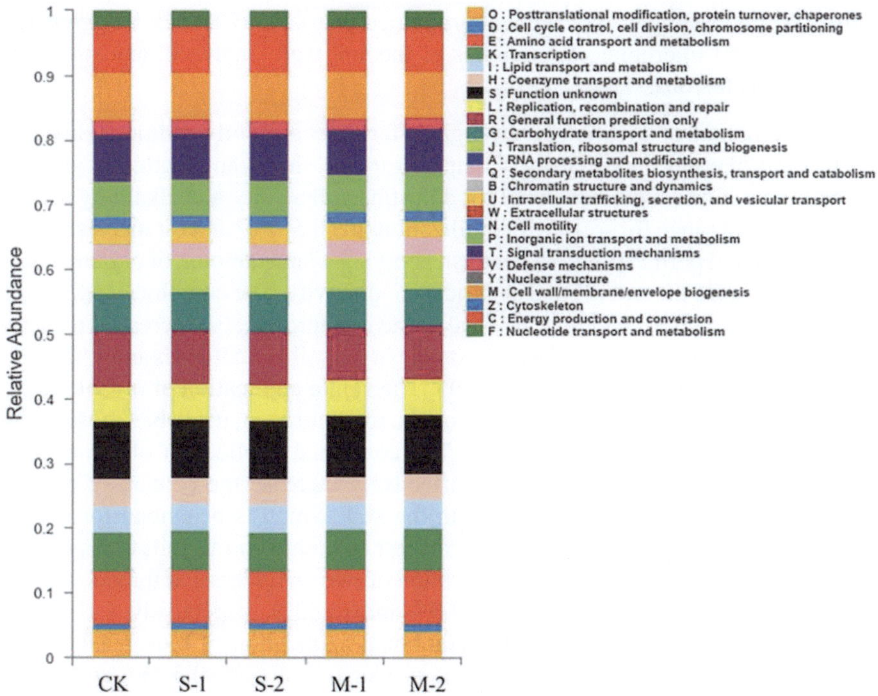

Fig. 5.14 Prediction of soil microbial functions for different fertilization treatments

protein; microorganisms that can decompose extracellular polysaccharides; and nitrogen-fixing and phosphorus-solubilizing bacteria, which is beneficial for the optimization of functional microbial groups in the soil.

(3) OM, humus, and large aggregates were positively correlated, while OM, humus, and aggregates were all negatively correlated with TDS. The families *Sphingomonadaceae* and *Xanthomonadaceae* were negatively correlated with TDS and positively correlated with OM, humus, and large aggregates. Conversely, *norank_c_Acidobacteria* was significantly negatively correlated with SOM, humus, and large aggregates, and positively correlated with TDS.

5.4 Chapter Summary

In this chapter, five treatments were investigated to study the amelioration effects of different fertilization regimes on greenhouse soil: no fertilizer, a single low-rate application of bio-organic fertilizer, a single high-rate application of bio-organic fertilizer, a combined application of bio-organic fertilizer and a food waste-derived conditioner, and a combined application of bio-organic fertilizer, a food waste conditioner, and a mineral-phytohormone conditioner. The effects of the five fertilization regimes on the basic physicochemical properties, nutrient content, soil aggregates, salt content, and microbial communities of the soil were analyzed, leading to the following conclusions:

(1) The combined application of bio-organic fertilizer and soil conditioner lowered the soil pH and bulk density. Compared to the single application of organic fertilizer, it was more beneficial for adjusting soil acidity and alkalinity toward the optimal range for soil fertility utilization (pH 6.5–7.5) and for increasing soil porosity and permeability. In contrast, applying a large amount of organic fertilizer increased the soil pH and electrical conductivity. The combined application of bio-organic fertilizer and the food waste conditioner also increased the total nitrogen and available phosphorus content in the soil to 1.59 times and 1.95 times that of the control group, respectively. The single application of organic fertilizer did not have a significant effect on the total nitrogen, available phosphorus, or available potassium in the soil. The combined application of bio-organic fertilizer and soil conditioner can effectively increase the content of organic matter and humus-like substances in the soil, which is beneficial for carbon sequestration and promoting the formation of humus in the soil. Compared to no fertilization, the combined application of two fertilizers and three fertilizers increased soil organic matter by 40.60% and 75.52%, respectively.
(2) Compared to no fertilization, the combined application of bio-organic fertilizer and soil conditioner reduced the proportion of 0.25–0.5 mm aggregates and increased the proportion of > 1 mm aggregates. The single application of organic fertilizer did not have a significant effect on the size of soil aggregates. After the combined application of bio-organic fertilizer and the food waste conditioner, the mean weight diameter and geometric mean diameter of the aggregates increased significantly, reaching 1.46 times and 1.58 times that of the no-fertilizer group, respectively. In general, the combined application of organic fertilizer and the food waste conditioner can enhance soil aggregation, promote the cementation of small-sized aggregates into large-sized aggregates, increase the content of large soil aggregates, promote the formation of large aggregates, and improve their stability, thereby improving soil structure.
(3) The main cations in the greenhouse soil were Ca^{2+}, K^+, and Na^+, and the main anions were NO_3^- and Cl^-. The combined application of bio-organic fertilizer and soil conditioner can significantly reduce the total soluble salt content in the soil, and can also decrease the concentrations of K^+, Na^+, and Ca^{2+}, lowering the proportion of cations in the total ions. This is beneficial for regulating the balance of anions and cations in the soil and reducing the toxic effects of single salts. Compared to the control group, the total soluble salt content decreased by 3.44% and 32.38% after the combined application of two and three fertilizers, respectively, effectively reducing the accumulation of salt in the surface soil. In contrast, after the application of a large amount of bio-organic fertilizer, a large amount of Ca^{2+} and Na^+ accumulated in the cultivated layer of the soil, and the salt content was 1.95 times that of the no-fertilizer group, which is not conducive to solving the problem of soil salinization.

(4) Fertilization did not have a significant effect on the diversity and richness of bacteria in the soil. However, the combined application of bio-organic fertilizer and soil conditioner had a relatively large impact on the soil microbial community. It increased the abundance of dominant microbial groups such as *Proteobacteria, Bacteroidetes*, and *Gemmatimonadetes*, while reducing the abundance of microorganisms such as *Acidobacteria*. It also increased the abundance of functional bacteria, including those that decompose macromolecules like cellulose, starch, and protein; microorganisms that can decompose extracellular polysaccharides; and nitrogen-fixing and phosphorus-solubilizing bacteria, which is beneficial for the optimization of functional groups in the soil.
(5) OM, humus, and large aggregates were positively correlated, while OM, humus, and aggregates were all negatively correlated with TDS. The families *Sphingomonadaceae* and *Xanthomonadaceae* were negatively correlated with TDS and positively correlated with OM, humus, and large aggregates. Conversely, *Blastocatellaceae_Subgroup_4_* and *norank_c_Acidobacteria* were significantly negatively correlated with OM, humus, and large aggregates, and positively correlated with TDS.

References

Chardon W, Oenema O, Del Castilho P (1997) Organic phosphorus in solutions and leachates from soils treated with animal slurries. J Environ Qual 26(2):372–378
Chen C, Xiao B, Yu Y (2009) Spatial variability of soil organic matter and pH and the correlation to available nutrients in the tea garden of southern Shaanxi. J Northwest A&F Univ (Nat Sci Ed) 01:190–196
He X, Xi B, Gao R, Zhang H, Dang Q, Li D, Huang C (2016) Insight into the composition and degradation potential of dissolved organic matter with different hydrophobicity in landfill leachates. Chemosphere 144:75–80
Hou J, Li M, Mao X, Hao Y, Ding J, Liu D, Xi B, Liu H (2017) Response of microbial community of organic-matter-impoverished arable soil to long-term application of soil conditioner derived from dynamic rapid fermentation of food waste. PLoS ONE 12(4):e0175715
Liu M, Wang C, Wang F (2019) Vermicompost and humic fertilizer improve coastal saline soil by regulating soil aggregates and the bacterial community. Arch Agron Soil Sci 65(3):281–293
Mbarki S, Cerdà A, Brestic M (2017) Vineyard compost supplemented with Trichoderma harzianum T78 improve saline soil quality. Land Degrad Dev 28(3):1028–1037
Qi R (2016) Responses of soil organic carbon mineralization under long-term fertilization regimes to temperature changes and cattle manure addition. Chinese Academy of Agricultural Sciences
Schulten H, Schnitzer M (1993) A state-of-the-art structural concept for humic substances. Naturwissenschaften 80(1):29–30
Tattini R, Gucci M, Coradeschi C, Everard J (2010) Growth, gas exchange and ion content in Olea europaea plants during salinity stress and subsequent relief. Physiol Plant 95:203–210
Wang C, Liu S, Zhang Y (2018) Bacterial communities and their predicted functions explain the sediment nitrogen changes along with submerged macrophyte restoration. Microb Ecol 76(3):625–636
Yang J (2018) Study on the salinity characteristics and effect of lowering soil salt concentration using carbon regulator. Yangzhou University

Yu H (2006) Nutrient state and characteristics of accumulation and translocation of salt in greenhouse soil profiles. Sichuan Agricultural University

Zhang C, Lin Y, Tian X (2017) Tobacco bacterial wilt suppression with biochar soil addition associates to improved soil physiochemical properties and increased rhizosphere bacteria abundance. Appl Soil Ecol 112:90–96

Zhang Y, Wang J, Shen M (2010) Effect of long-term fertilization on soil acidification in Taihu lake region, China. Acta Pedologica Sinica 47(3):465–472

MIX
Papier aus verantwortungsvollen Quellen
Paper from responsible sources
FSC® C105338

If you have any concerns about our products,
you can contact us on
ProductSafety@springernature.com

In case Publisher is established outside the EU,
the EU authorized representative is:
Springer Nature Customer Service Center GmbH
Europaplatz 3, 69115 Heidelberg, Germany

Printed by Libri Plureos GmbH
in Hamburg, Germany